REPRESENTATION THEORY
A CATEGORICAL APPROACH

Representation Theory

A Categorical Approach

Jan E. Grabowski

OpenBook Publishers

https://www.openbookpublishers.com

©2025 Jan E. Grabowski

ISBN Paperback: 978-1-80511-714-8
ISBN Hardback: 978-1-80511-715-5
ISBN PDF: 978-1-80511-716-2
DOI: https://doi.org/10.11647/OBP.0492

Cover image: Spectrum Patterns, September 27, 2024, under the Unsplash+ License, https://unsplash.com/illustrations/a-drawing-of-a-bunch-of-different-shapes-too-XlOUtGQ
Cover design: Jeevanjot Kaur Nagpal

Introduction

This book is an introduction to *representation theory*. Representation theory is complementary to *structure theory*, these being the two main goals within the study of abstract algebra. When we define a class of algebraic structures—such as groups, rings or fields—we can ask about their properties and constituent parts, then attempt to classify interesting subclasses. This is what we would call "structure theory".

In contrast, representation theory initially concentrates on one particular member of a class—a particular group, ring or field, say—and asks in what ways it occurs as the symmetries of other objects. The first example one usually meets is that of a group acting on sets. Many groups arise as the symmetries of a geometric shape, such as triangles, squares, tetrahedra, cubes or more complicated shapes.

If we concentrate on the symmetry group of a triangle, a group usually called S_3, we can ask "what other shapes (or more generally sets) also have S_3-symmetry?" An example would be a hexagon, as we can see by inscribing a triangle inside it. Then another way to phrase our question is "what is the representation theory of S_3?"

This book is intended for those who have learned linear algebra and some structure theory of groups or rings[1] and who would like to find out about representation theory.[2] Where needed, for example to establish notation, we will recall definitions and relevant results and signpost to other sources for more details.

We will also take a *categorical* approach. That is, we will start by introducing the notion of a category and related ideas, to give us the language we need to state and prove results in a way that aligns with current research in representation theory. We will use only a little category theory *per se* but will make frequent use of notions such as categories of modules, morphisms in those categories, functors and so on. This will give our presentation of some classical results a different flavour, complementing other (excellent) texts covering similar material, such as [EH], [ASS], [JL] or [Sch].

Inevitably, it is not possible to cover every topic in one book. We have chosen not to address the non-associative theory, that is, Lie algebras and their associated Lie groups. Part of the reason for this is that the flavour of the representation theory of Lie algebras is somewhat different. Another reason is that there are a number of very good extant books on this topic, such as [FH] and [Car]. Similarly, one has to stop somewhere with the top-

---

[1]In the UK higher education context, this would usually be in the first or second year of an undergraduate degree in mathematics.

[2]The advanced topics in Chapter 6 aim to take the reader to the point of being ready to read recent research papers in relevant areas of representation theory.

©2025 Jan E. Grabowski, CC BY-NC 4.0 https://doi.org/10.11647/OBP.0492.00

ics one has and so this volume does not cover character theory for finite groups; we recommend [JL] to those wishing to explore the representation theory of groups further.

Any text of this type inevitably draws on the work of many previous authors and is influenced by one's colleagues. I would like to offer general thanks to all of them, some of whom are cited within the text. A special mention goes to Michael Wemyss, however: we taught our first lecture courses in Oxford in 2008, mine on group theory and his on algebras and representation theory. This book reflects more than fifteen years of conversation between us on approaches to the pedagogy of abstract algebra, not least its attempt to say things in the most general way possible that the audience can appreciate.

I would particularly like to thank the reviewers of the drafts of this text, whose suggestions were extremely helpful and whose comments have immeasurably improved the whole book.

JEG, April 2025

Note to students

Some sections are marked with the symbol ☡, indicating that they are more advanced topics. Generally they can be omitted until such time as you are ready for them. In them, there will often be fewer details, including undefined terminology. When this happens, it is expected that you will use explicit references together with helpful on- or off-line sources to fill in the gaps.

Note to fellow educators

This text emerged from the author's desire to give a synthesized approach to teaching advanced algebra, resisting the common (often pragmatically motivated) separation of the representation theories of finite groups and algebras respectively. However, one might wish or need to teach these topics separately, as indeed the author had need to. I therefore offer the following recommendation of how to construct lecture courses of approximately twenty hours of lectures:

Representation Theory of Finite Groups: 1.1–1.5, 2, 3.1, 4, (optionally, 5.1), 5.2 (plus three to four hours on characters, if desired)

Representation Theory of Algebras: 1, 2, 3.2, 3.3, 4, (optionally, 5.1), 5.3

Judgement based on students' prior knowledge will determine the speed at which 1.2–1.4 need to be covered.

Contents

About the author

Jan Grabowski is an experienced researcher and educator in mathematics, with over twenty years' experience of both, and currently holds the position of Professor of Algebra at Lancaster University. Jan has a strong research track record of publications in algebra and related topics, both as a single author and collaboratively. He has taught courses at Oxford and Lancaster across a range of levels and has been teaching abstract algebra for the majority of this time, including covering aspects of the material in this book in courses at both institutions.

Jan has had recognition for his teaching, including a teaching prize at Oxford, obtaining a Postgraduate Certificate in Academic Practice and being awarded Senior Fellowship of the Higher Education Academy. He has championed innovation in teaching by presenting a mature exposition of 'how mathematicians really think about these things' as compared with other approaches that often defer more advanced ideas or techniques, rather than encouraging students to engage with challenging ideas early and repeatedly.

Personal website: https://www.maths.lancs.ac.uk/ grabowsj/

Chapter 1

Algebra

The aim of this first chapter is to gather the definitions and some fundamental properties of the main protagonists in our story: groups, rings, fields, vector spaces, algebras and quivers. We start with those you are more likely to be familiar with and you should move through these at the pace that suits you.

Some details will be omitted and we will not venture into areas that are not closely related to our main goal in this book. For example, we will not discuss factorization properties in rings, which you might otherwise find in a book covering ring theory. If you would like to explore abstract algebra further, the bibliography (p. 211) contains several good starting points at different levels, of which the most wide-ranging is [Alu].

1.1   Algebraic structures

The way we usually characterize *algebraic structures* is as sets with some data, most often *operations* or *distinguished elements*. We usually ask that the operations or distinguished elements have certain properties or that different operations or elements interact in a particular way. In principle there are many different types of algebraic structure but in practice some have much richer theories than others and more (interesting) examples "in nature".

An algebraic structure consisting of a set S and n pieces of data will be denoted by recording the information as tuples, e.g. $(S, d_1, d_2, \ldots, d_n)$. Sometimes we will use different letters for the set S, to help us remember which type of algebraic structure we are thinking about, and we might use established or more suggestive notation for the operations or elements, such as $+, \times, 0$ or 1.

Some important properties an operation might have include being *associative*, meaning that $a * (b * c) = (a * b) * c$ for all $a, b, c \in S$, or being *commutative*, meaning that $a * b = b * a$ for all $a, b \in S$. We will also regularly encounter the notion of an *identity element* for an operation $*$, which is an element e such that $e * a = a = a * e$ for all $a \in S$. When we have an identity element, it is also possible to define *inverse* elements: we say that b is a (two-sided) inverse for a with respect to $*$ if $a * b = e = b * a$. It is a helpful exercise to show that identity elements and inverses are unique if they exist.

©2025 Jan E. Grabowski, CC BY-NC 4.0 https://doi.org/10.11647/OBP.0492.01

Here are some examples of algebraic structures you will encounter throughout this book.

Definition 1.1.1. A group is an algebraic structure $(G, *)$ with $*$ an associative binary operation, having an identity element e and such that every element of G has an inverse with respect to $*$.

Definition 1.1.2. A group $(G, *)$ is called Abelian if $*$ is commutative.

The order of a group $(G, *)$ is the cardinality of G, denoted $|G|$ and we say that $(G, *)$ is finite if the set G is.

Definition 1.1.3. A ring is an algebraic structure $(R, +, \times)$ such that $(R, +)$ is an Abelian group, $\times$ is associative and we have the distributive laws

$$a \times (b + c) = (a \times b) + (a \times c)$$

and

$$(a + b) \times c = (a \times c) + (b \times c)$$

for all $a, b, c \in R$. The identity element for the group $(R, +)$ will be denoted 0_R.

The two equations given here are called the first and second (or left and right) distributive laws. We could be more general and say that $\times$ distributes over $+$ from the left if $a \times (b + c) = (a \times b) + (a \times c)$, and correspondingly for the right-handed version. Notice that these are not symmetric in $+$ and $\times$.

Looking carefully, you may notice that this definition of a ring does not necessarily match with that that you have seen before. This is because we have not included in the definition the requirement for a ring to have a *multiplicative identity* (i.e. an identity element with respect to $\times$), also sometimes called a unit. We will call a ring with an identified multiplicative identity for $\times$ a *unital ring*. To emphasize that we might not have a multiplicative identity, we will sometimes say that $(R, +, \times)$ satisfying only Definition 1.1.3 is a (*not necessarily unital*) ring. If one can prove that the ring cannot have a multiplicative identity, we say that it is *non-unital*.

Definition 1.1.4. A field is an algebraic structure $(F, +, \times)$ such that $(F, +, \times)$ is a ring, $\times$ is commutative and has an identity $1_F \neq 0_F$ and every element of $F \setminus \{0_F\}$ has an inverse with respect to $\times$.

Notice that the definition of a field $(F, +, \times)$ could also be written equivalently as saying that $(F, +)$ is an Abelian group with identity 0_F, $(F \setminus \{0_F\}, \times)$ is an Abelian group and that $+$ and $\times$ satisfy the distributive laws.

From now on, we will prefer the symbol $\mathbb{K}$ for a field; this will avoid clashes of notation as we have another concept for which we will prefer F.[1]

We see from these definitions that rings are groups with extra data, and fields are rings with further extra data. So every field is in particular a ring, and every ring is a group.

Groups, rings and fields appear in various different contexts and the following list of examples is not in any way comprehensive. Rather, they are examples that should be somewhat familiar to you following a first course in abstract algebra.

Number systems

- The rational numbers $\mathbb{Q}$, the real numbers $\mathbb{R}$ and the complex numbers $\mathbb{C}$ are fields.

- The set of integers $\mathbb{Z}$ with the usual operations is a unital ring (but not a field); in many ways, $\mathbb{Z}$ is the prototypical (unital) ring and many ring-theoretic questions are inspired by the behaviour of $\mathbb{Z}$.

- The set of even integers $2\mathbb{Z}$ with the operations inherited from $\mathbb{Z}$ is a non-unital ring.

- Note that $\mathbb{N} = \{0, 1, 2, \dots\}$ is *not* a group or a ring with respect to the usual addition and multiplication operations. Note too that we choose the convention that $0 \in \mathbb{N}$.

- For each n, the set of integers modulo n, $\mathbb{Z}_n$ (with addition and multiplication modulo n) is a unital ring and is a field if and only if n is prime.

- The set of polynomials in n variables with coefficients from a ring R, $R[x_1, \dots, x_n]$, is again a ring, with respect to the usual addition and multiplication of polynomials. This includes examples such as $\mathbb{R}[x]$, $\mathbb{C}[x, y]$ and so on.

- Since $\mathbb{Q}$, $\mathbb{R}$ and $\mathbb{C}$ are fields, as we said in the remarks after the definition of a field, we have associated groups $\mathbb{Q}^\times = \mathbb{Q} \setminus \{0\}$, $\mathbb{R}^\times = \mathbb{R} \setminus \{0\}$ and $\mathbb{C}^\times = \mathbb{C} \setminus \{0\}$ with respect to multiplication.

  More generally, the set of invertible elements in a ring R forms a group under multiplication (called the group of units, $R^\times$). For a field $\mathbb{K}$, we have $\mathbb{K}^\times = \mathbb{K} \setminus \{0_\mathbb{K}\}$, because every non-zero element is invertible.

---

[1]The use of a form of the letter K here is common among representation theorists, deriving from the German terminology "Körper" ("body") due originally to Dedekind.

Symmetry groups

- Natural examples of symmetry groups are the set of rotations and re-flections of a square, an n-gon, a circle or other 2-dimensional figures, as well as rotations and reflections of a cube, tetrahedron, etc.

  A particularly important example that occurs often is the *dihedral group* D_{2n} (of order $2n$), defined to be the group of isometries of $\mathbb{R}^2$ pre-serving a regular n-gon. The dihedral group D_{2n} is generated by two elements a and b and has the following presentation:

  $$D_{2n} = \langle a, b \mid a^n = e, \; b^2 = e, \; a^i b = ba^{-i} \; \forall \, i \in \mathbb{Z} \rangle$$

  It follows that $D_{2n} = \{a^i b^j \mid 0 \le i \le n - 1, \; 0 \le j \le 1\}$ and hence that D_{2n} has $2n$ elements. Here, the elements a^i are the n rotations of the n-gon and the elements $a^i b$ are the n reflections.

- The "symmetries of a set" S are given by the set of bijections
  $\mathrm{Bij}(S) \overset{\text{def}}{=} \{f \colon S \to S \mid f \text{ is a bijection}\}$. This set satisfies all the condi-tions to be a group with respect to composition of functions.

  Permutations are exactly bijections from the set $\{1, \dots, n\}$ to itself. These groups are called the symmetric groups and we denote them by S_n.

  They (and the symmetry groups of geometric objects) are usually not Abelian, in contrast to all the previous examples.

Matrices

- We can add $m \times n$ matrices and multiply $n \times n$ matrices, we think we know what the zero and identity matrices are and we know that sometimes we can invert matrices. So it should not be a surprise that sets of matrices can have either group or ring structures, though they are almost never fields.

  More precisely, $\mathrm{M}_n(\mathbb{Z})$, $\mathrm{M}_n(\mathbb{Q})$, $\mathrm{M}_n(\mathbb{R})$ and $\mathrm{M}_n(\mathbb{C})$ are all rings; in fact $\mathrm{M}_n(R)$ is one for any ring R. We call these *matrix rings*. (In linear algebra, we usually work over a field, because we want to be able to divide by scalars as well as multiply by them.)

- Inside any matrix ring over a field $\mathbb{K}$, the subset of invertible matrices (those with non-zero determinant) forms a group, the group of units, as defined above. We have a special name and notation for the group of units of matrix rings: we call them *general linear groups* and write $\mathrm{GL}_n(\mathbb{K})$. The group operation is matrix multiplication.

Sets of functions

- The set of functions from $\mathbb{R}$ to $\mathbb{R}$, denoted $\mathcal{F}(\mathbb{R}, \mathbb{R})$, is a ring with respect to pointwise addition and multiplication, that is, for functions f, g and $x \in \mathbb{R}$, $(f + g)(x) = f(x) + g(x)$ and $(fg)(x) = f(x)g(x)$.

- We can generalize this to functions from any set Y to $\mathbb{R}$ (denoted $\mathcal{F}(Y, \mathbb{R})$) or from any set Y to $\mathbb{C}$ (denoted $\mathcal{F}(Y, \mathbb{C})$) or indeed from any set Y to any ring R (denoted $\mathcal{F}(Y, R)$). Each of these is a ring, essentially because the codomain ($\mathbb{R}$, $\mathbb{C}$ or R) is.

This list might give the impression that there are a lot more rings than groups. This is both misleading and complicated. Every ring is an Abelian group under addition and not every group is Abelian, but on the other hand, a given Abelian group might support multiple ring structures. It just so happens that the examples you are familiar with are mostly rings or fields, but there are certainly plenty of groups too.

One important observation about the above list is that lots of these examples, notably the symmetric groups, the matrix rings and general linear groups, are naturally *symmetries* of something. The description of symmetry groups we gave says this explicitly. One of the main results in linear algebra is that matrices are the same thing as linear transformations of vector spaces, and hence $\mathrm{GL}_n(\mathbb{K})$ is the symmetry group of $\mathbb{K}^n$, the n-dimensional vector space over $\mathbb{K}$. This idea is what leads us to study representation theory.

In the next two sections, we include foundational definitions and properties of groups, rings and fields that will be used (often tacitly) in the main body of this book. We will give less commentary than elsewhere in the book and also omit proofs: if you have not seen them before, we recommend at least trying to prove them yourself. Depending on your background, you may wish to review this material in more or less detail. However, you should resume no later than Section 1.4, since there we cover important prerequisites for our later work.

Throughout this text, boxes such as these will appear occasionally. They will encourage you to use computer algebra systems to help you with computing examples of the theory being discussed. The main motivation for this is to give you a tool to explore larger examples than are feasible by hand, so that you get a better sense of the general case than you might from very small examples.

In the main, this will focus on doing the heavy lifting of linear algebra, e.g. computing eigenvalues and eigenvectors, applying elementary matrix transformations etc. But many computer algebra systems can do much more, so we are only dipping our toes in the water and we encourage you to explore the wealth of material in the online tutorials and manuals to find out more.

We have also chosen to concentrate on the system SageMath ([Sag]). SageMath integrates a number of free open-source packages, is based on Python and has a handy browser-based interface available called SageMath-Cell (`https://sagecell.sagemath.org/`). Each of these comes with a number of advantages for us, hence our choice.

However, you may have access to a different computer algebra system and prefer to use this. You will find that your system can probably do everything we will ask SageMath to do, once you have translated the syntax appropriately. Generally speaking we will be using in-built functions, so this will largely be a matter of finding the corresponding function name in the manual. Be aware, however, that different systems do sometimes have different conventions and so the outputs may not exactly match what is included here from SageMath. This is what is known in the business as a "Learning Opportunity"; understanding why the answers look different will probably help you understand the mathematics better.

In this section, we will introduce SageMath and its syntax. Our approach will be to give code snippets to get you started, not to write a textbook on using SageMath. You will almost certainly find it helpful to use the Sage Tutorial (`https://doc.sagemath.org/html/en/tutorial/index.html`) alongside this book, to find more in depth guidance and information. Note that we have tested code given here via SageMath-Cell, which is using version 10.6 at the time of writing. It is possible that changes have happened in between then and your reading this book, in which case some further investigation may be needed[a] to identify appropriate changes to the syntax.

The first basic operation is assigning names and values. Typing

```
1  n = 10
```

into a SageMathCell and pressing "Evaluate" returns nothing. This is because what we have done is asked SageMath to assign the value 10 to the variable n, which (hopefully) it did. If we want to see output, we have to ask for it: entering

```
2  n
```

on a new line and pressing "Evaluate" prints 10, as we wanted. Setting another value, say $m = 5$, we can ask true/false questions as follows:

```
3  m = 5
4  m == n
```

evaluates to `False` and we can ask multiple further questions at once by adding

```
5  (m <= n, 7 > m)
```

to obtain (`True`,`True`). Note that we can do multiple calculations on one line, separated by semicolons ; but Evaluate just returns the value of the last of these. So

```
6   m  <= n;  7 > m
```

is valid syntax but just returns `True`. If we had asked

```
6   m  >= n;  7 > m
```

we would have been told `True`, even through the first expression evaluates to `False`.

To summarize, = assigns values, == compares for equality and <, <=, > and >= have their expected meanings as "less than", "less than or equal to" and so on. Be aware that SageMath objects have (dynamic) types and comparisons like this may fail due to incomparability of type rather than value.

---

^aThe author would welcome all feedback but especially that on non-functioning code, so that this can be corrected for the online edition and future print editions.

1.2 Groups

We begin by recording some very basic propositions about groups.

Proposition 1.2.1. *Let* $(G, *)$ *be a group.*

(i) *The identity element of G, e, is unique.*

(ii) *Every element of G has a unique inverse with respect to $*$.*

(iii) *(Cancellation) For $a, b, c \in G$, if either $a * b = a * c$ or $b * a = c * a$ then $b = c$.*

(iv) *For $a, b \in G$, the equation $a * x = b$ has the unique solution $x = a^{-1} * b$.*

(v) *For $a, b \in G$, the equation $x * a = b$ has the unique solution $x = b * a^{-1}$.*

(vi) *If $a * b = e$ then $b = a^{-1}$ and $a = b^{-1}$.*

(vii) *For $a, b \in G$, $(a * b)^{-1} = b^{-1} * a^{-1}$.*

The following is adapted from the "Group Theory and Sage" tutorial, to illustrate (vii).

```
1   G = SymmetricGroup(4)
2   a = G([(1,2,3)])
3   b = G([(2,3,4)])
```

```
4   a * b
5   (a * b)^(-1)
6   (b^(-1)) * (a^(-1))
7   (a^(-1)) * (b^(-1))
```

Evaluating

- after line 4 tells us that a * b is equal to (1 3)(2 4),

- after line 5 that this is also (a * b)^(-1),

- after line 6, that this is also equal to (b^(-1)) * (a^(-1)) and

- after line 7, that this is not equal to (a^(-1)) * (b^(-1)), which is (1 2)(3 4).

Unexplained mathematical notation will be covered in more detail shortly, but hopefully you see what is happening in SageMath.

Note that SageMath *computes in specific examples*, rather than proving things in general from axioms. Other types of computer system, such as the proof assistant Lean, do the latter.

Corollary 1.2.2. *Let $(G, *)$ be a group and let $a \in G$. The functions $L_a : G \to G$ defined by $L_a(g) = a * g$ and $R_a : G \to G$ defined by $R_a(g) = g * a$ are bijections.*

Proposition 1.2.3. *Let $(G, *)$ and $(H, \circ)$ be groups. Then the Cartesian product $G \times H$ is a group with respect to the binary operation $\bullet$ defined by*

$$(g_1, h_1) \bullet (g_2, h_2) = (g_1 * g_2, h_1 \circ h_2)$$

for all $g_1, g_2 \in G$, $h_1, h_2 \in H$.

1.2.1 Subgroups

In examining some class of algebraic objects, typically the first question one asks is "how do I make more of these?". Gluing them together, as in the Cartesian product construction above, is one way but the other is to look inside and find *subobjects*. This not only gives us more objects from our class but also a close and important relationship between them: if A is a subset of B, there is a natural injective function from A to B encoding the inclusion.

Subobjects are typically defined by saying that they are given by finding a subset such that the operations restrict to the elements of the subset and retain the same properties.

Definition 1.2.4. Let $(G, *)$ be a group. A subset H of G is said to be a subgroup if $(H, *)$ is a group. If this is the case, we write $H \leq G$.

From the definition, we see that "H is a subgroup of G" means not just that "H is a group with respect to *some* operation", but it means that H is a group with respect to the *same* operation as G.

Remarks 1.2.5.

(a) It is immediate from this definition that every subgroup of an Abelian group is Abelian: if $a*b = b*a$ for all $a, b \in G$, then certainly $a*b = b*a$ for all $a, b \in H \subseteq G$.

(b) Remember that $* \colon G \times G \to G$ is a function sending pairs of elements of G to an element in G. In order for $(H, *)$ to be a group, when we restrict this function to $H \times H$, to obtain $*|_{H \times H} \colon H \times H \to G$, we have to have $\operatorname{Im} *|_{H \times H} \subseteq H$ so that there is a well-defined function $*|_{H \times H}^{H} \colon H \times H \to H$, restricting the domain to $H \times H$ and the codomain to H, and $(H, *|_{H \times H}^{H})$ is a group. Unsurprisingly, we tend to brush this under the carpet a bit and just write $(H, *)$.

Examples 1.2.6.

(a) $(\mathbb{Z}, +)$ is a subgroup of $(\mathbb{Q}, +)$, and both are subgroups of $(\mathbb{R}, +)$.

(b) $(\mathbb{Q}^*, \times)$ is a subgroup of $(\mathbb{R}^*, \times)$.

(c) $(\mathbb{Q}^*, \times)$ is *not* a subgroup of $(\mathbb{Q}, +)$, since the two binary operations are different.

(d) For $n \in \mathbb{N}$ let $\zeta = e^{2\pi i/n} \in \mathbb{C}$. The set $C_n = \{\zeta^k \mid k \in \mathbb{Z}\}$ is a subgroup of $(\mathbb{C}^*, \times)$. Notice that $|C_n| = n$ (since $e^{2\pi i} = 1$), so that C_n is a finite group, while $\mathbb{C}^*$ is certainly not.

(e) In the group $\operatorname{GL}_n(\mathbb{R})$ (where the operation is matrix multiplication), the set of matrices with determinant 1 is a subgroup; this follows from properties of determinants. This subgroup has a special name: it is known as the special linear group, $\operatorname{SL}_n(\mathbb{R})$. (As before, one may replace $\mathbb{R}$ by any field $\mathbb{K}$ and obtain $\operatorname{SL}_n(\mathbb{K})$.)

(f) We can consider S_{n-1} as a subgroup of S_n: if $\sigma \in S_{n-1}$ then we think of σ as a permutation of $\{1, 2, \ldots, n\}$, by setting $\sigma(n) = n$. More generally, the set of permutations fixing every element of some subset of $\{1, 2, \ldots, n\}$ is a subgroup: if two permutations fix some i so does their product, and if a permutation fixes i, so does its inverse.

(g) In any group $(G, *)$ with identity element e, the subsets $\{e\}$ (the trivial subgroup) and G are subgroups. If H is a subgroup of G, then when $H \neq \{e\}$ we say H is non-trivial and if $H \neq G$ we say H is proper.

In principle, to show that a subset H of a group $(G, *)$ is itself a group— that is, to check the definition of a subgroup above—one should show that H has a well-defined binary operation that is associative, has an identity element and where every element has an inverse. However this is a lot to check and in fact one can get away with doing rather less.

Proposition 1.2.7 (Subgroup test). *Let H be a subset of a group $(G, *)$ and denote the identity element of G by e_G. Then H is a subgroup of G if and only if the following three conditions are satisfied:*

(SG1) $e_G \in H$;

(SG2) *for all $h_1, h_2 \in H$, we have $h_1 * h_2 \in H$;*

(SG3) *for all $h \in H$, we have $h^{-1} \in H$.*

Let $(G, *)$ be a group and consider the set

$$Z(G) = \{z \in G \mid z * g = g * z \text{ for all } g \in G\}.$$

Then $Z(G)$ is a subgroup of G and $Z(G)$ is Abelian. This subgroup is important enough to be given a name, the *centre* of G.

Problem 1. Using the subgroup test, show that $Z(G)$ is a subgroup of G.

If I is a set, we say that a collection of sets $\mathcal{A}$ is an I-indexed family of sets if there is a one-to-one correspondence (i.e. a bijection) between the sets in the collection $\mathcal{A}$ and I. Then, fixing such a bijection, we may write $\mathcal{A} = \{A_i \mid i \in I\}$, i.e. denoting by A_i the member of $\mathcal{A}$ corresponding to I.

The intersection of the sets in $\mathcal{A}$ is $\bigcap_{i \in I} A_i = \{x \mid (\forall i \in I)(x \in A_i)\}$. If $\mathcal{H} = \{H_i \mid i \in I\}$ is an I-indexed family of subgroups of a fixed group G, then $\bigcap_{i \in I} H_i$ is also a subgroup of G. Indeed, we do not even need to index the subgroups.

Proposition 1.2.8. *For any collection $\mathcal{H}$ of subgroups of G, we may define $\bigcap \mathcal{H} = \{x \mid (\forall H \in \mathcal{H})(x \in H)\}$. Then $\bigcap \mathcal{H}$ is a subgroup of G.*

Remark 1.2.9. The union of subgroups need not be a subgroup: let $H_1 = 2\mathbb{Z}$, $H_2 = 3\mathbb{Z}$ be subgroups of $(\mathbb{Z}, +)$. Then $2, 3 \in H_1 \cup H_2$ but $2 + 3 = 5 \notin H_1 \cup H_2$.

Example 1.2.10. In $(\mathbb{Z}, +)$, the intersection of $2\mathbb{Z}$ and $3\mathbb{Z}$ is the set of integers that are divisible by 2 and divisible by 3, and these are precisely the integers divisible by 6. So $2\mathbb{Z} \cap 3\mathbb{Z} = 6\mathbb{Z}$.

More generally, $m\mathbb{Z} \cap n\mathbb{Z} = \text{lcm}(m, n)\mathbb{Z}$ and

$$m_1\mathbb{Z} \cap m_2\mathbb{Z} \cap \cdots \cap m_r\mathbb{Z} = \text{lcm}(m_1, m_2, \ldots, m_r)\mathbb{Z}.$$

The collection $\mathcal{H} = \{m\mathbb{Z} \mid m \in \mathbb{N}\}$ is a $\mathbb{N}$-indexed family of subgroups of $\mathbb{Z}$. The intersection $\bigcap_{m \in \mathbb{N}} m\mathbb{Z}$ consists of all integers that are divisible by *every* $m \in \mathbb{N}$. Only 0 has this property, so $\bigcap_{m \in \mathbb{N}} m\mathbb{Z} = \{0\}$ is trivial.

Here are some elementary commands for constructing $\mathbb{Z}$ and its subgroups; since $\mathbb{Z}$ is an infinite group, SageMath is limited in what it can compute. Note too that SageMath constructs a *multiplicative* Abelian group, i.e. $Z = \langle f \rangle$, which is isomorphic to $(\mathbb{Z}, +)$ but we have to write expressions like f^2.

```
1  Z = AbelianGroup (1)
2  Z.order ()
3  f = Z.gen ()
4  mZ = Z.subgroup ([f^2])
5  nZ = Z.subgroup ([f^3])
6  mZ == nZ
7  mZ.is_isomorphic (nZ)
```

Again, we suggest evaluating after each line to see the different outputs.

Sometimes we have a subset S of a group G that is not a subgroup, but where we would like to find a subgroup of G that contains S. Obviously we could take all of G but this won't tell us much: we really want the *smallest* subgroup of G containing S.

"Smallest" here might mean "smallest size", but for infinite sets it is better to think of smallest as "smallest under inclusions of sets". But we know how to find the smallest set: it is the intersection. (The intersection is a subset of every set in the collection and is the unique smallest such under inclusion.) This leads us to make the following definition.

Definition 1.2.11. Let G be a group and let $S \subseteq G$ be a subset of G. Let $\mathcal{S} = \{H \leq G \mid S \subseteq H\}$ be the family of subgroups of G that contain S. The intersection $\bigcap \mathcal{S}$ is a subgroup of G containing S, called the subgroup of G generated by S. We write $\langle S \rangle = \bigcap \mathcal{S}$.

Notice that by Proposition 1.2.8, $\langle S \rangle$ is a subgroup of G. It is also the smallest subgroup having the desired property, namely that it contains S. If S is actually a subgroup already, then $\langle S \rangle = S$.

Example 1.2.12. In $(\mathbb{Z}, +)$, we have $\langle m \rangle = m\mathbb{Z}$ (and in particular $\langle 1 \rangle = \mathbb{Z}$) and $\langle \{m, n\} \rangle = \mathrm{hcf}(m, n)\mathbb{Z}$, where hcf denotes the highest common factor (also known as the greatest common divisor).

This key example leads us to the following definition.

Definition 1.2.13. We say a group G is *cyclic* if there exists $g \in G$ such that $\langle g \rangle = G$.

That is, cyclic groups are 1-generator, i.e. can be generated by a single element. With a little work, one can show that a finite cyclic group is necessarily isomorphic to the group of integers modulo n, $\mathbb{Z}_n$, for some n and any infinite cyclic group is isomorphic to $\mathbb{Z}$.

We will often use a "generic" cyclic group C_n, given by

$$C_n = \langle g \mid g^n = e \rangle$$

Here we create C_5 in SageMath, find out that its order is 5, multiply some elements, get a list of all its elements and ask how many subgroups it has.

```
1  C5 = AbelianGroup([5])
2  C5.order()
3  f = C5.gen()
4  e = C5.identity()
5  f*e*f
6  C5.list()
7  C5.number_of_subgroups()
```

To simplify notation, if $S = \{g_1, \ldots, g_r\}$, we will write $\langle g_1, \ldots, g_r \rangle$ rather than $\langle \{g_1, \ldots, g_r\} \rangle$ for the subgroup generated by S. So for example, we would write $\langle m, n \rangle = \mathrm{hcf}(m, n)\mathbb{Z}$.

The above description of the subgroup generated by S as an intersection is good from the point of view of proving that it is actually a subgroup, but bad from the point of view of knowing what $\langle S \rangle$ looks like as a subset of G.

So we are now going to try to tie down a more precise description of the subgroup generated by S, as the set of elements of G which can be "expressed in terms of" the elements of S. For example, if $x, y \in S$ (and hence in $\langle S \rangle$ since $S \subseteq \langle S \rangle$) then $xy^2x^{-1}y^2x^3y^{-1}x$ is expressed in terms of x and y and is in $\langle S \rangle$, by repeated application of the subgroup test.

The following proposition expresses this idea in more formal language.

Proposition 1.2.14. *Let S be a subset of G. Let H be the set of all elements of G of the form $s_1^{a_1} s_2^{a_2} \cdots s_r^{a_r}$, where r is a non-negative integer, $s_1, \ldots, s_r \in S$ and $a_1, \ldots, a_r \in \mathbb{Z}$. (When $r = 0$, this is the identity element of G.)*
Then H is a subgroup of G and $H = \langle S \rangle$.

1.2.2   Permutations

Recall that permutations of n elements are bijections from the set $\{1, 2, \ldots, n\}$ to itself and that we write S_n for the set of all such bijections. However, it is not very convenient to use this definition for calculations, so in examples, we will use *disjoint cycle notation* for permutations.

Definition 1.2.15. Let $r \leq n$ and let $a_1, a_2, \ldots, a_r$ be a list of distinct elements of $\{1, 2, \ldots, n\}$. Let $\pi \in S_n$ be the permutation defined by

$$\pi(a_i) = a_{i+1} \qquad \text{for } 1 \leq i \leq r - 1$$
$$\pi(a_r) = a_1$$
$$\pi(a) = a \qquad \text{for } a \notin \{a_1, a_2, \ldots, a_r\}$$

We will call π a cycle of length r, or an r-cycle, and we will write $(a_1 \, a_2 \, \cdots \, a_r)$.

For example, the permutation that in two-line notation is $\left(\begin{smallmatrix} 1 & 2 & 3 \\ 2 & 3 & 1 \end{smallmatrix}\right)$ is a 3-cycle, namely (2 3 1). Similarly, the permutation $\left(\begin{smallmatrix} 1 & 2 & 3 & 4 & 5 \\ 1 & 4 & 5 & 3 & 2 \end{smallmatrix}\right)$ is a 4-cycle, namely (2 4 3 5); notice that the numbers in a cycle do not have to come in increasing order and that 1 is fixed so does not appear in this cycle.

Remark 1.2.16. We note the following properties of cycles:

(a) An r-cycle can be written in r different ways, since

$$(a_1 \, a_2 \, \cdots \, a_r) = (a_2 \, a_3 \, \cdots \, a_r \, a_1) = \cdots = (a_r \, a_1 \, \cdots \, a_{r-1}).$$

We usually write cycles with the smallest element of $\{a_1, a_2, \ldots, a_r\}$ first.

(b) The identity permutation ι is represented by any 1-cycle; this looks a bit odd but the seeming ambiguity will be resolved shortly (by disjoint cycle notation).

(c) We can compose two cycles by 'feeding in' elements from the right in turn; for example, if $\pi = (1 \ 2 \ 5)$ and $\sigma = (2 \ 3 \ 4)$ then $\pi \circ \sigma = (1 \ 2 \ 5)(2 \ 3 \ 4)$ maps

$$1 \mapsto 1 \mapsto 2, \quad 2 \mapsto 3 \mapsto 3, \quad 3 \mapsto 4 \mapsto 4, \quad 4 \mapsto 2 \mapsto 5, \quad 5 \mapsto 5 \mapsto 1.$$

We often say this as "1 goes to 1, 1 goes to 2; 2 goes to 3, 3 goes to 3" and so on. That is, we start with 1 and write "(1 ". Then we see where the first and then the second permutation send it, and write the partial permutation "(1 2 "; next it would be "(1 2 3 " and so on, until we reach an element that is sent to 1, at which point we close the cycle. So $\pi \circ \sigma = (1 \ 2 \ 3 \ 4 \ 5)$.

(d) The inverse of a cycle is obtained by simply writing its elements in reverse order; for example if π is as above then $\pi^{-1} = (5 \ 2 \ 1) = (1 \ 5 \ 2)$. This makes sense because if π sends 2 to 5, then π^{-1} should send 5 to 2, and so on. (We can also check this by computing $(1 \ 2 \ 5)(5 \ 2 \ 1)$.)

(e) If we compose an r-cycle with itself r times, the result will be the identity permutation.

(f) If $\pi = (a_1\ a_2\ \cdots\ a_r)$ and $\sigma = (b_1\ b_2\ \cdots\ b_s)$ are cycles such that the sets $\{a_i \mid 1 \leq i \leq r\}$ and $\{b_j \mid 1 \leq j \leq s\}$ are disjoint, then $\pi \circ \sigma = \sigma \circ \pi$. We can see this because the elements moved by π and σ are different, so it does not matter whether π or σ is performed first.

As an example, $\pi = (1\ 2\ 3)$ and $\sigma = (4\ 5\ 6\ 7)$ are permutations in S_8, are *disjoint* cycles and

$$\pi \circ \sigma = (1\ 2\ 3)(4\ 5\ 6\ 7)$$
$$= \begin{pmatrix} 1\ 2\ 3\ 4\ 5\ 6\ 7\ 8 \\ 2\ 3\ 1\ 5\ 6\ 7\ 4\ 8 \end{pmatrix}$$
$$= (4\ 5\ 6\ 7)(1\ 2\ 3) = \sigma \circ \pi.$$

Now not all permutations are cycles. The permutation $\begin{pmatrix} 1\ 2\ 3\ 4\ 5\ 6\ 7\ 8 \\ 2\ 3\ 1\ 5\ 6\ 7\ 4\ 8 \end{pmatrix}$ above is not a cycle, for example. However, it is a product of two cycles which move disjoint sets of points; this suggests how we may extend the idea of cycles to cover all permutations.

Definition 1.2.17. Let $\pi = (a_1\ a_2\ \cdots\ a_r)$ and $\sigma = (b_1\ b_2\ \cdots\ b_s)$ be cycles in S_n. If the sets $\{a_1, a_2, \ldots, a_r\}$ and $\{b_1, b_2, \ldots, b_s\}$ are disjoint, we say that π and σ are disjoint cycles. This terminology is extended in the obvious way to sets of more than two cycles.

For the permutation $\alpha = \begin{pmatrix} 1\ 2\ 3\ 4\ 5\ 6\ 7\ 8 \\ 2\ 3\ 1\ 5\ 6\ 7\ 4\ 8 \end{pmatrix}$ above, it is easy to recover the cycles $(1\ 2\ 3)$ and $(4\ 5\ 6\ 7)$ that it is a product of. Based on the above remark, we can write it equally well as any of the following:

$$\alpha = \begin{pmatrix} 1\ 2\ 3\ 4\ 5\ 6\ 7\ 8 \\ 2\ 3\ 1\ 5\ 6\ 7\ 4\ 8 \end{pmatrix}$$
$$= (1\ 2\ 3)(4\ 5\ 6\ 7)$$
$$= (4\ 5\ 6\ 7)(1\ 2\ 3)$$
$$= (1\ 2\ 3)(4\ 5\ 6\ 7)(8)$$
$$= (2\ 3\ 1)(4\ 5\ 6\ 7)(8)$$

or indeed in various other equivalent ways. The following theorem shows that the behaviour observed in this example is typical.

Theorem 1.2.18. *Every permutation in S_n can be written as a product of disjoint cycles; moreover this expression is unique up to*

(i) *the order in which the cycles occur,*

(ii) *the different ways of writing each cycle (rotating the cycle), and*

(iii) *the presence or absence of 1-cycles.*

Definition 1.2.19. A permutation which is written as a product of disjoint cycles is said to be in disjoint cycle notation; if all 1-cycles are included, it is said to be in full cycle notation.

In disjoint cycle notation, we often leave out the 1-cycles. Then the identity permutation ι in disjoint cycle notation has no cycles at all; this is why we give it a notation, ι, of its own.

Example 1.2.20. Let $\pi = \left(\begin{smallmatrix} 1 & 2 & 3 & 4 & 5 & 6 & 7 & 8 & 9 \\ 3 & 8 & 6 & 1 & 7 & 4 & 9 & 2 & 5 \end{smallmatrix}\right)$ and $\sigma = \left(\begin{smallmatrix} 1 & 2 & 3 & 4 & 5 & 6 & 7 & 8 & 9 \\ 1 & 5 & 6 & 9 & 7 & 2 & 3 & 8 & 4 \end{smallmatrix}\right)$; then the expressions for $\pi, \sigma, \pi\sigma$ and π^{-1} in full cycle notation are as follows:

$$\pi = (1\ 3\ 6\ 4)(2\ 8)(5\ 7\ 9),$$

$$\sigma = (2\ 5\ 7\ 3\ 6)(4\ 9) = (2\ 5\ 7\ 3\ 6)(4\ 9)(1)(8),$$

$$\pi\sigma = (1\ 3\ 6\ 4)(2\ 8)(5\ 7\ 9)(2\ 5\ 7\ 3\ 6)(4\ 9)$$
$$= (1\ 3\ 4\ 5\ 9)(2\ 7\ 6\ 8),$$

$$\pi^{-1} = (4\ 6\ 3\ 1)(8\ 2)(9\ 7\ 5) = (1\ 4\ 6\ 3)(2\ 8)(5\ 9\ 7).$$

Let us do these calculations in SageMath:

```
1  G = SymmetricGroup(9)
2  pi = G("(1,3,6,4) (2,8) (5,7,9)")
3  sigma = G("(2,5,7,3,6) (4,9)")
4  sigma * pi
5  pi^-1
```

Warning: SageMath composes permutations in the opposite way to our notation. That is, by $\pi\sigma$ we mean do σ then π, but SageMath writes this as sigma * pi, with the permutation to be applied first on the left.

Recall that composing an r-cycle with itself r times gives the identity permutation ι.

Definition 1.2.21. Let $\pi \in S_n$. The order $o(\pi)$ of π is the minimal $r \in \mathbb{N}$ such that $\pi^r = \iota$.

Then one particular advantage of disjoint cycle notation is that it is easy to calculate the order of a permutation written this way.

Proposition 1.2.22. *If $\pi = \gamma_1\gamma_2\cdots\gamma_k$, where the γ_i are disjoint cycles, then the order of π is the lowest common multiple of the lengths of the cycles γ_i.*

The smallest non-trivial cycles have a special name.

Definition 1.2.23. A 2-cycle is called a transposition.

Theorem 1.2.24. *If $n > 1$, every permutation in S_n can be written as a product of transpositions.*

We next briefly recap the notion of the sign of a permutation.

Let $\pi \in S_n$ and suppose that $\pi = \gamma_1 \gamma_2 \ldots \gamma_k$ with the γ_i *disjoint* cycles. Set

$$\nu(\pi) = \sum_{i=1}^{k} (|\gamma_i| - 1),$$

where $|\gamma_i|$ denotes the length of the cycle γ_i. Note that $\nu(\pi)$ is well-defined, although this requires some care (in particular, cycles of length 1 contribute 0 to the sum, and so may be ignored). Note that π^{-1} has the same cycle lengths as π, so we have $\nu(\pi^{-1}) = \nu(\pi)$.

Clearly $\nu(\pi) = 1$ if and only if π is a transposition. In fact we may interpret $\nu(\pi)$ as follows: if all possible 1-cycles are included in the expression $\pi = \gamma_1 \gamma_2 \ldots \gamma_k$, then $\sum_{i=1}^{k} |\gamma_i| = n$; thus $\nu(\pi) = n - k$, i.e. $\nu(\pi)$ is the difference between n and the number of cycles in the full cycle notation.

Definition 1.2.25. Given $\pi \in S_n$, the sign of π is defined by

$$\mathrm{sign}(\pi) = (-1)^{\nu(\pi)}.$$

Thus $\mathrm{sign}(\pi) = \pm 1$; if $\mathrm{sign}(\pi) = 1$ we call π an even permutation, while if $\mathrm{sign}(\pi) = -1$ we call π an odd permutation.

Note that because $\nu(\pi^{-1}) = \nu(\pi)$, we have $\mathrm{sign}(\pi^{-1}) = \mathrm{sign}(\pi)$.

We now consider how the sign of a permutation is affected by multiplication by a transposition.

Lemma 1.2.26. *Let $\pi \in S_n$ and let $\tau \in S_n$ be a transposition. Then*

$$\mathrm{sign}(\pi\tau) = (-1) \cdot \mathrm{sign}(\pi) = -\mathrm{sign}(\pi).$$

Corollary 1.2.27. *If $\tau_1, \ldots, \tau_r \in S_n$ are transpositions, then*

$$\mathrm{sign}(\tau_1 \cdots \tau_r) = (-1)^r.$$

Theorem 1.2.28. *Let $\pi, \sigma \in S_n$. Then $\mathrm{sign}(\pi)\,\mathrm{sign}(\sigma) = \mathrm{sign}(\pi\sigma)$.*

This result shows that if we define a function $\Sigma \colon S_n \to \{1, -1\}$ by setting $\Sigma(\pi) = \mathrm{sign}(\pi)$ for all $\pi \in S_n$, then Σ is a group homomorphism, as defined below.

Continuing the above,

```
6  pi.sign()
7  sigma.sign()
8  rho = sigma * pi
9  rho.sign() == sigma.sign() * pi.sign()
```

returns True, as expected.

Notice too that if we take the product of two even permutations, we obtain another even permutation ($1 \times 1 = 1$), so the subset of S_n consisting of all the even permutations is closed with respect to the group operation. This particular subgroup of S_n is important enough to have its own name.

Definition 1.2.29. The subgroup of S_n consisting of the even permutations is called the alternating group of degree n and is denoted A_n.

In SageMath, the construction is

```
1  A7 = AlternatingGroup(7)
2  A7.order()
```

It can be shown that any element of A_n can be written as a product of 3-cycles (the 'smallest' elements of A_n, since every transposition is odd).

Remark 1.2.30. Another way to describe the sign of a permutation is via certain linear maps, or equivalently their matrices. For a given permutation $\sigma \in S_n$, let f_σ be the linear map defined by acting on the standard basis $\{e_i \mid 1 \leq i \leq n\}$ of $\mathbb{R}^n$ as $f_\sigma(e_i) = e_{\sigma(i)}$; that is, f_σ permutes the basis of $\mathbb{R}^n$ using the permutation σ. In the language we will adopt later, this is the *linearization* of the action of σ (see Section 3.1.3, Definition 5.2.3 and Example 5.2.10).

Then the matrix of f_σ is A_σ, the $n \times n$ matrix with 1 in the $(\sigma(i), i)$ entry $(1 \leq i \leq n)$ and 0 everywhere else. Alternatively, $A_\sigma = \sum_{i=1}^n e_{\sigma(i),i}$, where e_{ji} is the $n \times n$ matrix with 1 in the (j, i) position and 0 everywhere else.

We have $A_\sigma A_\tau e_i = A_\sigma e_{\tau(i)} = e_{\sigma\tau(i)} = A_{\sigma\tau}e_i$, for all i, so $A_\sigma A_\tau = A_{\sigma\tau}$.

What this shows is that $\sigma \mapsto A_\sigma$ defines a function $S_n \to \mathrm{GL}_n(\mathbb{R})$ that respects multiplication. In this setting, the sign of a permutation σ is just $\det A_\sigma$. The matrices A_σ are called permutation matrices or monomial matrices.

As an explicit example, let $\sigma = (1\ 2\ 3\ 4)$ and $\tau = (1\ 3\ 2)$ be permutations in S_4. We have

$$A_\sigma = \begin{pmatrix} 0 & 0 & 0 & 1 \\ 1 & 0 & 0 & 0 \\ 0 & 1 & 0 & 0 \\ 0 & 0 & 1 & 0 \end{pmatrix}, \qquad A_\tau = \begin{pmatrix} 0 & 1 & 0 & 0 \\ 0 & 0 & 1 & 0 \\ 1 & 0 & 0 & 0 \\ 0 & 0 & 0 & 1 \end{pmatrix},$$

$$A_{\sigma\tau} = A_{(1\ 4)} = \begin{pmatrix} 0 & 0 & 0 & 1 \\ 0 & 1 & 0 & 0 \\ 0 & 0 & 1 & 0 \\ 1 & 0 & 0 & 0 \end{pmatrix} \qquad \text{and}$$

$$A_\sigma A_\tau = \begin{pmatrix} 0 & 0 & 0 & 1 \\ 1 & 0 & 0 & 0 \\ 0 & 1 & 0 & 0 \\ 0 & 0 & 1 & 0 \end{pmatrix}\begin{pmatrix} 0 & 1 & 0 & 0 \\ 0 & 0 & 1 & 0 \\ 1 & 0 & 0 & 0 \\ 0 & 0 & 0 & 1 \end{pmatrix} = \begin{pmatrix} 0 & 0 & 0 & 1 \\ 0 & 1 & 0 & 0 \\ 0 & 0 & 1 & 0 \\ 1 & 0 & 0 & 0 \end{pmatrix} = A_{\sigma\tau}.$$

1.2.3 Cosets

Throughout this section, we will consider a group G and a subgroup H of G. Now we will mostly write the group operation in G by concatenation: if $g, h \in G$, gh will denote the element obtained from g and h (in that order) under the binary operation of the group. The main exception will be in groups where the binary operation is addition (such as $(\mathbb{Z}, +)$), where we will write $g + h$.

Definition 1.2.31. For any $g \in G$, the set $gH = \{gh \mid h \in H\}$ will be called the left coset of H in G determined by g.

Example 1.2.32. In S_3, let

$$H = \langle (1\ 2\ 3) \rangle = \{\iota, (1\ 2\ 3), (1\ 3\ 2)\}.$$

The coset $(1\ 2)H$ in S_3 is

$$\begin{aligned} \{ \quad & (1\ 2)\iota = (1\ 2), \\ & (1\ 2)(1\ 2\ 3) = (1)(2\ 3) = (2\ 3), \\ & (1\ 2)(1\ 3\ 2) = (1\ 3)(2) = (1\ 3) \quad \}. \end{aligned}$$

SageMath can compute cosets, but because it composes permutations the other way round to us, our left cosets are its right cosets:

```
1  S3 = SymmetricGroup(3)
2  h = S3("(1,2,3)")
3  H = S3.subgroup([h])
4  S3.cosets(H,side="right")
```

The output is of two lists, each containing three elements, and one of which is [(2,3), (1,2), (1,3)], matching the above example. The other coset, [(), (1,2,3), (1,3,2)] is H.

First of all, we see that $gH \subseteq G$. Since $e \overset{\text{def}}{=} e_G \in H$, we have $g = ge \in gH$. Indeed, every element of gH is, by definition, a "multiple" of g by an element of H; the left coset gH is precisely all the elements of G that can be written as gh for some $h \in H$.

If $H = \{e, h_1, h_2, \ldots, h_r\}$ is finite, this is clear: $gH = \{g, gh_1, gh_2, \ldots, gh_r\}$ and by Proposition 1.2.1(iii) (the cancellation property for groups) $gh_i = gh_j$ would imply $h_i = h_j$, so the elements $g, gh_1, \ldots, gh_r$ are distinct. In particular, $|gH| = |H|$. It is natural to think of gH as the "translations by g" of the elements of H.

In fact, the claims of the above paragraph still hold if H is infinite, if we work a little more abstractly.

Lemma 1.2.33. *Let H be a subgroup of G. For any $g \in G$, there is a bijection $L_g \colon H \to gH$ given by $L_g(h) = gh$.*

So any two cosets of H have the same size. Notice too that $eH = H$ so H is a coset of itself (the "translation" of H that does nothing).

However it is very much *not* true that if $g_1 \neq g_2$ then $g_1 H \neq g_2 H$; shortly we will see the correct condition for when two cosets are equal. This condition will also tell us when a coset can be a subgroup (sneak preview: only the "trivial" coset $eH = H$ is).

First, let us look at a very important example.

Example 1.2.34. We consider $(\mathbb{Z}, +)$ and its (cyclic) subgroup $4\mathbb{Z}$. Since we are working in an additive group, we write the group operation in our coset notation, so the cosets of $4\mathbb{Z}$ are $r + 4\mathbb{Z} = \{r + m \mid m \in 4\mathbb{Z}\} = \{r + 4n \mid n \in \mathbb{Z}\}$.

We see that the left cosets of $4\mathbb{Z}$ are *precisely* the congruence classes of integers modulo 4. The congruence class modulo 4 containing 3, for example, is the set of integers $\{\ldots, -9, -5, -1, 3, 7, 11, \ldots\}$, which is exactly the set of integers of the form $3 + 4n$, namely $\{3 + 4n \mid n \in \mathbb{Z}\} = 3 + 4\mathbb{Z}$. So cosets in particular describe congruence classes of integers.

Notice that every integer belongs to some left coset and that no integer belongs to more than one left coset. This is because the congruence classes modulo n for some fixed n *partition* $\mathbb{Z}$, since congruence modulo n is an equivalence relation. The following theorem is a vast generalisation of this result, which corresponds to the specific group $(\mathbb{Z}, +)$.

Theorem 1.2.35. *Let G be a group and let H be a subgroup of G.*

(i) *The relation $\mathcal{R}_H$ on G defined by*

$$g \,\mathcal{R}_H\, k \iff g^{-1}k \in H$$

is an equivalence relation.

(ii) *The equivalence classes for $\mathcal{R}_H$ are precisely the left cosets gH, $g \in G$.*

Corollary 1.2.36. *For H a subgroup of G, the set of left cosets $L_H = \{gH \mid g \in G\}$ is a partition of G.*

That is, for $g_1, g_2 \in G$, $g_1 H \cap g_2 H$ is either empty or $g_1 H = g_2 H$ (and the latter happens if and only if $g_1^{-1} g_2 \in H$), and furthermore $\bigcup_{g \in G} gH = G$.

Corollary 1.2.37. *For a subgroup H of G, $gH = kH$ if and only if $g^{-1}k \in H$.*

The number of cosets of a given subgroup H of a group G then provides some measure of the relative size of H in G.

Definition 1.2.38. Let G be a group and let H be a subgroup of G. The index of H in G is defined to be the cardinality of the set of left cosets of H in G. We denote the index of H in G by $|G : H|$.

For the example above,

```
libgap.Index(S3, H)
```

returns 2 as expected. Note the use of "libgap.", to tell SageMath to use functionality provided by GAP (one of several systems SageMath can outsource its calculations to).

Example 1.2.39. For $n \geq 2$, the set of even permutations A_n in S_n has index $|S_n : A_n| = 2$. By Theorem 1.2.28, if $\pi \notin A_n$ and $\tau = (1\ 2)$ then

$$\mathrm{sign}(\tau\pi) = \mathrm{sign}(\tau)\,\mathrm{sign}(\pi) = (-1)(-1) = 1$$

so $\tau\pi \in A_n$ and $\pi = \tau(\tau\pi) \in \tau A_n$. That is, for any $\pi \in S_n$, either $\pi \in A_n$ or $\pi \in \tau A_n$ and the two cosets A_n and τA_n of A_n in S_n partition S_n. Therefore the index of A_n in S_n is at most 2. But transpositions are odd, so $\tau \notin A_n$ and so there are at least two cosets of A_n and the index is exactly 2.

For $n = 1$, $A_1 = S_1 = \{\iota\}$ so the index is 1.

When we think about the set of left cosets of a subgroup, it is very natural to want a complete list of the cosets, without repetitions. But we have already seen that a coset of H in G has $|H|$ representatives, so clearly such a list will not be unique in general. Still, we can simply make some choice.

Definition 1.2.40. Let H be a subgroup of a group G. A transversal for H in G is a set $T_H \subseteq G$ such that for every coset gH, there exists a unique element $t_g \in T_H$ such that $gH = t_g H$.

That is, a transversal is a complete set of representatives of the left cosets of H in G, such that any two distinct cosets have distinct representatives.

If we are looking at a specific group and subgroup, there might be one or more "natural" choices of transversal. This happens in particular for the integers under addition and the subgroup $n\mathbb{Z}$. As we saw for $4\mathbb{Z}$ above, the cosets correspond to the different integers modulo n, so a natural transversal for $n\mathbb{Z}$ in $(\mathbb{Z}, +)$ is

$$T_{n\mathbb{Z}} = \{r \mid 0 \leq r \leq n - 1\}.$$

Explicitly, for $n = 4$, we can choose $T_{4\mathbb{Z}} = \{0, 1, 2, 3\}$ as the set of left cosets is $L_{4\mathbb{Z}} = \{0 + 4\mathbb{Z}, 1 + 4\mathbb{Z}, 2 + 4\mathbb{Z}, 3 + 4\mathbb{Z}\}$.

Other choices are also perfectly valid: the sets $\{1, 2, 3, 4\}$, $\{0, 5, 18, 1003\}$ and $\{-3, -2, -1, 0\}$ are all transversals for $4\mathbb{Z}$. However, it is natural to choose $\{0, 1, 2, 3\}$ as these are the minimal positive representatives, and we will usually make this choice and take $T_{n\mathbb{Z}}$ as above.

Note however that $\{-2, -1, 0, 1, 2\}$ is another natural choice for a transversal of $5\mathbb{Z}$, and a similar choice can be made for $(2m + 1)\mathbb{Z}$ (i.e. odd n). Since this works less well for even n, it is less commonly used, until one wants to consider $p\mathbb{Z}$ for p prime, when in all but one case p is odd.

You will probably have noticed that we have been saying "left cosets" and that the definition of gH is asymmetric. Unsurprisingly, one can make the corresponding definition of a right coset.

Definition 1.2.41. For any $g \in G$, the set $Hg = \{hg \mid h \in H\}$ will be called the right coset of H in G determined by g.

Analogously to Definition 1.2.38, define the right index of H in G to be the number of right cosets of H in G and denote it by $|H : G|$.

In general the left cosets of a subgroup are not the same as the right cosets; certain special conditions have to hold for this to be the case.

Example 1.2.42. In S_4, let

$$K = \langle (1\ 2\ 3\ 4) \rangle = \{\iota, (1\ 2\ 3\ 4), (1\ 3)(2\ 4), (1\ 4\ 3\ 2)\}.$$

The right coset $K(1\ 2)$ in S_4 is

$$\begin{aligned}
\{ \quad & \iota(1\ 2) = (1\ 2), \\
& (1\ 2\ 3\ 4)(1\ 2) = (1\ 3\ 4), \\
& (1\ 3)(2\ 4)(1\ 2) = (1\ 4\ 2\ 3) \\
& (1\ 4\ 3\ 2)(1\ 2) = (2\ 4\ 3) \qquad \}
\end{aligned}$$

and the left coset $(1\ 2)K$ is $\{(1\ 2), (2\ 3\ 4), (1\ 3\ 2\ 4), (1\ 4\ 3)\}$ which we see is not equal to $K(1\ 2)$.

SageMath can do this for us too:

```
1   S4 = SymmetricGroup(4)
2   k = S4("(1,2,3,4)")
3   K = S4.subgroup([k])
4   S4.cosets(K,side="left")
5   S4.cosets(K,side="right")
6   S4.cosets(K,side="left") == S4.cosets(K,side="
        right")
```

Here, evaluating after line 4 and then after line 5, we can find the left and right cosets containing $(1\ 2)$ and compare them with the above (remembering that we have to interchange left and right). Line 6 confirms that the set of left cosets is not the same as the set of right cosets.

So we must be careful to specify whether we mean left cosets or right cosets. While we need to make a choice, some key properties are independent of which choice we make. In particular, the size of a left or a right coset is the same: every coset, whether left or right, has the same size as H (there is a natural right-handed version of Lemma 1.2.33).

A little more work also shows that the number of left cosets of H in G is equal to the number of right cosets of H in G; that is, $|G : H| = |H : G|$. In fact, this is why we just say "index" and not "left index" or "right index". This is justified by the following exercise.

Problem 2. Let H be a subgroup of G. Let $L_H = \{gH \mid g \in G\}$ and $R_H = \{Hg \mid g \in G\}$ be the sets of left and right cosets of G respectively. Prove that the function $\psi \colon L_H \to R_H$, $\psi(gH) = Hg^{-1}$ is a bijection.

This function ψ is a function between two sets: in general, the set of cosets of a subgroup has no algebraic structure.

Remark 1.2.43. Subgroups such that $gH = Hg$ for all $g \in G$ are very important, so much so that they have a special name. They are called normal subgroups and we will examine them and their importance in detail in the following sections.

But, as a glimpse of what is to come, we will see that the set of (left) cosets of a *normal* subgroup can be given an algebraic structure. Specifically, the group operation descends to a group operation on the set of cosets: if $gH = Hg$ for all $g \in G$, we can define a binary operation on cosets by

$$(gH)(kH) = g(Hk)H = g(kH)H = gkH^2 = gkH.$$

Continuing from above,

```
7  K.is_normal()
```

returns False.

The fundamental theorem involving cosets is the following.

Theorem 1.2.44 (Lagrange's theorem). *Let G be a finite group and H a subgroup of G. Then $|G| = |G : H||H|$.*

Corollary 1.2.45. *For any subgroup H of G, both $|H|$ and $|G : H|$ divide $|G|$.*

Conversely, if m does not divide $|G|$, G cannot have a subgroup of order m.

Remark 1.2.46. As written above and in its usual presentation in a first course, Lagrange's theorem applies to finite groups. However a version of the result is also true for infinite groups, as follows:

Problem 3. Let L_H denote the set of left cosets of H in G. Choose T_H a transversal for H, that is, $T_H \subseteq G$ such that for all $gH \in L_H$, there exists a unique $t_g \in T_H$ such that $gH = t_g H$ (this was Definition 1.2.40). Then for all $g \in G$, there exists a unique $h_g \in H$ such that $g = t_g h_g$.

Prove that the function $\varphi \colon G \to L_H \times H$, $\varphi(g) = (t_g H, h_g)$ is well-defined and a bijection.

1.2.4 Homomorphisms

When we have two groups G and H and we want to relate the group structure on G to that on H, to compare properties between them, the right mathematical thing to do is to start with a function $\varphi \colon G \to H$. This gives us a relationship between the underlying sets but this need not relate the group structures: we need φ to be compatible with the binary operations on G and H.

A function with this extra property will be called a (group) *homomorphism*, where "homo-" means "same" (in the sense of "consistent") and the suffix "morphism" indicates that the overall meaning is "the same structure", or perhaps more mathematically accurately, "compatible structure".

This is one of the fundamental concepts behind the use of category theory in advanced algebra in general and representation theory in particular.

This leads us to the following definition.

Definition 1.2.47. Let $(G, *)$ and $(H, \circ)$ be groups. A function $\varphi \colon G \to H$ is called a group homomorphism if

$$\varphi(g_1 * g_2) = \varphi(g_1) \circ \varphi(g_2)$$

for all $g_1, g_2 \in G$.

Notice in particular that the operation in G, $*$, is being used on the left-hand side of this equation (on $g_1, g_2 \in G$), and that the operation in H, $\circ$ is being used on the right-hand side (on $\varphi(g_1), \varphi(g_2) \in H$). This is an instance when it is helpful to write the group operations explicitly.

Note that the identity function $\mathrm{id}_G \colon G \to G$, $\mathrm{id}_G(g) = g$ is a group homomorphism.

First, let us deal with some elementary properties of homomorphisms.

Proposition 1.2.48. Let $\varphi \colon G \to H$ be a group homomorphism. Then

(i) $\varphi(e_G) = e_H$;

(ii) $\varphi(g^{-1}) = \varphi(g)^{-1}$ for all $g \in G$;

(iii) $\varphi(g^n) = \varphi(g)^n$ for all $g \in G$, $n \in \mathbb{Z}$ and

(iv) $o(\varphi(g)) \mid o(g)$ for all $g \in G$.

The composition of two functions is again a function and we would expect that if two composable functions preserved group structure, their composition would too. Indeed this is the case.

Proposition 1.2.49. Let $\varphi \colon G \to H$ and $\sigma \colon H \to K$ be group homomorphisms. Then $\sigma \circ \varphi \colon G \to K$ is a group homomorphism.

Examples 1.2.50.

(a) The function $\varphi\colon \mathbb{Z} \to \mathbb{Z}$, $\varphi(n) = 2n$ for all $n \in \mathbb{Z}$ is a group homomorphism from $(\mathbb{Z}, +)$ to itself:

$$\varphi(m + n) = 2(m + n) = 2m + 2n = \varphi(m) + \varphi(n)$$

for all $m, n \in \mathbb{Z}$. (Note the additive notation.)

(b) The function $\varphi\colon \mathbb{R}^* \to \mathbb{R}^*$, $\varphi(x) = x^2$ for all $x \in \mathbb{R}^* = (\mathbb{R} \setminus \{0\}, \times)$ is a group homomorphism since

$$\varphi(xy) = (xy)^2 = x^2 y^2 = \varphi(x)\varphi(y)$$

for all $x, y \in \mathbb{R}^*$.

(c) The function $\varphi\colon \mathbb{R} \to \mathbb{C}^*$, $\varphi(x) = e^{2\pi i x}$ for all $x \in \mathbb{R}$ is a group homomorphism from $(\mathbb{R}, +)$ to $(\mathbb{C} \setminus \{0\}, \times)$ since

$$\varphi(x + y) = e^{2\pi i(x+y)} = e^{2\pi i x + 2\pi i y} = e^{2\pi i x} e^{2\pi i y} = \varphi(x)\varphi(y)$$

for all $x, y \in \mathbb{R}$. (Note the two different notations for the group operations!)

(d) Recall from Remark 1.2.30 that we have a function $\varphi\colon S_n \to \mathrm{GL}_n(\mathbb{R})$, $\varphi(\sigma) = A_\sigma$ where

$$(A_\sigma)_{ij} = \begin{cases} 1 & \text{if } \sigma(j) = i \\ 0 & \text{otherwise} \end{cases}$$

is the permutation matrix associated to σ.

Then $A_{\sigma\tau} = A_\sigma A_\tau$, where the left-hand side involves the composition of permutations and the right-hand side is the usual multiplication of matrices. (The definition of A_σ, with a 1 when $\sigma(j) = i$ might seem to be the "wrong way round"; however if we made the more natural-looking definition with $\sigma(i) = j$, we would have an order reversal in the products.) It follows that $\varphi\colon S_n \to \mathrm{GL}_n(\mathbb{R})$, $\varphi(\sigma) = A_\sigma$ is a group homomorphism. For an explicit example, look back to Remark 1.2.30.

Problem 4. Let $(G, *)$ and $(H, \circ)$ be groups with identity elements e_G and e_H respectively. Show that the function $\varphi\colon G \to H$ defined by $\varphi(g) = e_H$ for all $g \in G$ is a group homomorphism. We call this the *trivial group homomorphism*.

Recall that certain sorts of functions are special, namely injective, surjective and bijective functions; also, a function is bijective if and only if it is invertible. If a group homomorphism has one of these properties, we (sometimes) use a special name for it. So an injective group homomorphism is also known as a monomorphism and a surjective group homomorphism is called an epimorphism. This comes from the wider category theory context, but in the setting of groups, being mono is equivalent to being injective and being epi is the same as being surjective.

Examples 1.2.51.

(a) The homomorphism $\varphi\colon \mathbb{Z} \to \mathbb{Z}$, $\varphi(n) = 2n$ for all $n \in \mathbb{Z}$ is injective but not surjective:

$$\varphi(m) = \varphi(n) \quad \Rightarrow \quad 2m = 2n \quad \Rightarrow \quad m = n$$

but $3 \notin \operatorname{Im} \varphi$.

(b) The homomorphism $\varphi\colon \mathbb{R}^* \to \mathbb{R}^*$, $\varphi(x) = x^2$ for all $x \in \mathbb{R}^*$ is not injective or surjective: $\varphi(-1) = 1 = \varphi(1)$ and $-1 \notin \operatorname{Im} \varphi$.

(c) The homomorphism $\varphi\colon \mathbb{R} \to \mathbb{C}^*$, $\varphi(x) = e^{2\pi i x}$ for all $x \in \mathbb{R}$ is not injective and not surjective:

$$\varphi(1) = e^{2\pi i} = 1 = \varphi(0)$$

and $\operatorname{Im} \varphi = \{z \in \mathbb{C} \mid \|z\| = 1\} \subsetneq \mathbb{C}^*$.

(d) The homomorphism $\varphi\colon S_n \to \mathrm{GL}_n(\mathbb{R})$, $\varphi(\sigma) = A_\sigma$ is injective but not surjective: given A_σ, we can easily recover σ (uniquely) but we see that $2I_n \notin \operatorname{Im} \varphi$.

(e) The function $\Sigma\colon S_n \to \{1, -1\}$, $\Sigma(\pi) = \operatorname{sign}(\pi)$ for all $\pi \in S_n$ is a homomorphism by Theorem 1.2.28, where $\{-1, 1\}$ is a group under multiplication.

Then Σ is not injective if $n \geq 3$ (there are typically many different permutations with sign 1; indeed these are the even permutations belonging to A_n, of which there are $n!/2$) but Σ is surjective if $n \geq 2$, since then there exist both even and odd permutations.

Definition 1.2.52. A group homomorphism $\varphi\colon G \to H$ is called a group isomorphism if there exists a group homomorphism $\psi\colon H \to G$ such that $\psi \circ \varphi = \operatorname{id}_G$ and $\varphi \circ \psi = \operatorname{id}_H$.

Here, the part "iso-" indicates equality, i.e. stronger "sameness" than "homo-" suggests in "homomorphism". Note that the identity homomorphism $\operatorname{id}_G\colon G \to G$, $\operatorname{id}_G(g) = g$ is its own inverse, so is a group isomorphism. This is no surprise: every group has exactly the same group structure as itself.

A group isomorphism is a particular kind of homomorphism—a function with certain properties. We can use these to talk about how two groups might be related, as follows.

Definition 1.2.53. We say that two groups G and H are isomorphic if and only if there exists a group isomorphism $\varphi\colon G \to H$. If G and H are isomorphic, we often write $G \cong H$, for short.

Lemma 1.2.54. *Let G and H be groups. Then the following are are equivalent:*

(i) *G and H are isomorphic, i.e. there exist group homomorphisms $\varphi\colon G \to H$ and $\psi\colon H \to G$ such that $\psi \circ \varphi = \mathrm{id}_G$ and $\varphi \circ \psi = \mathrm{id}_H$; and*

(ii) *there exists a bijective group homomorphism $\varphi\colon G \to H$.*

Indeed, we see from a careful examination of the proof that there is an equivalent condition

(iii) there exist functions $\varphi\colon G \to H$ and $\psi\colon H \to G$ such that $\psi \circ \varphi = \mathrm{id}_G$ and $\varphi \circ \psi = \mathrm{id}_H$ and *either* φ or ψ is a group homomorphism.

That is, if we know that a homomorphism is invertible, it is not necessary to check separately that the inverse is also a group homomorphism: this holds automatically. In particular, the inverse of an isomorphism is an isomorphism.

Since a group isomorphism $\varphi\colon G \to H$ is a bijection, we must have that $|G| = |H|$, i.e. isomorphic groups have the same order. The converse is most definitely false in general but the contrapositive tells us that if two groups have different orders, they cannot be isomorphic.

This leads us to seek *invariants* that help us identify when two given groups are isomorphic or not. If we can prove that a particular property is preserved under isomorphism, then if one group has the property but the other does not, they cannot be isomorphic.

Examples of invariants are the order of the group (as above), the set of natural numbers that are the orders of the elements of the group (see the next example), being Abelian and being cyclic.

Example 1.2.55. The groups C_6 and S_3 are not isomorphic. They have the same order ($|C_6| = 6 = 3! = |S_3|$) but C_6 is Abelian and S_3 is not. Indeed, C_6 is cyclic and S_3 is not (by considering cycle types, we see that S_3 has no elements of order 6).

```
1  S3 = SymmetricGroup(3)
2  C6 = CyclicPermutationGroup(6)
3  [C6.is_abelian(),S3.is_abelian(),S3.is_isomorphic(
       C6)]
```

returns [True,False,False].

However, sharing some properties is not (usually) enough to prove that the groups *are* isomorphic: almost always you will need to find an explicit isomorphism between them. That said, if two groups share lots of properties, this might reasonably lead you to conjecture that they are isomorphic; this would not be a proof, though.

Some combinations of properties are strong enough to conclude isomorphism, though.

Proposition 1.2.56. *Let G and H be cyclic groups. Then G is isomorphic to H if and only if they have the same order, $|G| = |H|$.*

Note that neither the statement nor the proof assume that the cyclic groups are finite. But by definition, an infinite cyclic group is countable ($|G| = |\mathbb{Z}| = |\mathbb{N}|$), so any two infinite cyclic groups have the same order and are therefore isomorphic.

Previously we noted that the identity map provides an isomorphism of a group with itself. In fact, a group can have other self-isomorphisms and it is often important to know how many. In some sense, these are the symmetries of the group (which itself may be the symmetries of something!).

Definition 1.2.57. Let G be a group. A group isomorphism $\varphi\colon G \to G$ is called an automorphism of G.

Lemma 1.2.58. *Let G be a group. The set of automorphisms of G forms a group under composition.*

We will see a far-reaching generalization of this lemma later, but for now just note that it justifies the following definition.

Definition 1.2.59. Let G be a group. The group of automorphisms of G, $\text{Aut}_{\textbf{Grp}}(G) = \{\varphi\colon G \to G \mid \varphi \text{ is an isomorphism}\}$ is called the automorphism group of G.

Example 1.2.60. One can show that $\text{Aut}_{\textbf{Grp}}(C_n) \cong (\mathbb{Z}_n^\times, \times)$, the group of units of the ring of integers modulo n. In particular, we have $\text{Aut}_{\textbf{Grp}}(C_2) = \{e\}$ and $\text{Aut}_{\textbf{Grp}}(C_3) \cong C_2$.

We will not justify these claims here: finding the automorphism group of even a small group usually requires more sophisticated technology than we currently have to hand. We will simply note that the above has an important relationship with number theory: $n \mapsto |\text{Aut}_{\textbf{Grp}}(C_n)|$ is called the Euler totient function.[2]

Instead, we will move on and look at some important subsets of a group associated to homomorphisms.

---

[2] Also known as Euler's phi function, since it is often denoted $\varphi(n)$.

1.2.5   Kernels and images

We already said that injective and surjective homomorphisms are particularly important; those that are both are very special, being isomorphisms. However most homomorphisms are neither injective nor surjective, so we would like a way to measure how far from being injective or surjective a given homomorphism is.

This is done by means of the kernel and image of the homomorphism. The image is just the image of the function: recall that being surjective precisely means that the image is all of the codomain. So the size of the image measures how close the map is to being surjective—the larger, the better.

The kernel is a subset of the domain whose size measures how close to being injective the map is: if the kernel is as small as possible, then the map is injective.

Definition 1.2.61. Let $\varphi\colon G \to H$ be a group homomorphism. The kernel of φ is defined to be the subset of G given by

$$\operatorname{Ker}\varphi = \{g \in G \mid \varphi(g) = e_H\}.$$

Definition 1.2.62. Let $\varphi\colon G \to H$ be a group homomorphism. The image of φ is defined to be the subset of H given by

$$\operatorname{Im}\varphi = \{h \in H \mid (\exists g \in G)(\varphi(g) = h)\}.$$

Notice that the definition of the kernel would not be possible without the identity element e_H and that the image is exactly the image of φ as a function. Be very clear in your mind that $\operatorname{Ker}\varphi$ is a subset of the domain, G, and $\operatorname{Im}\varphi$ is a subset of the codomain, H.

You might find the following picture helpful:

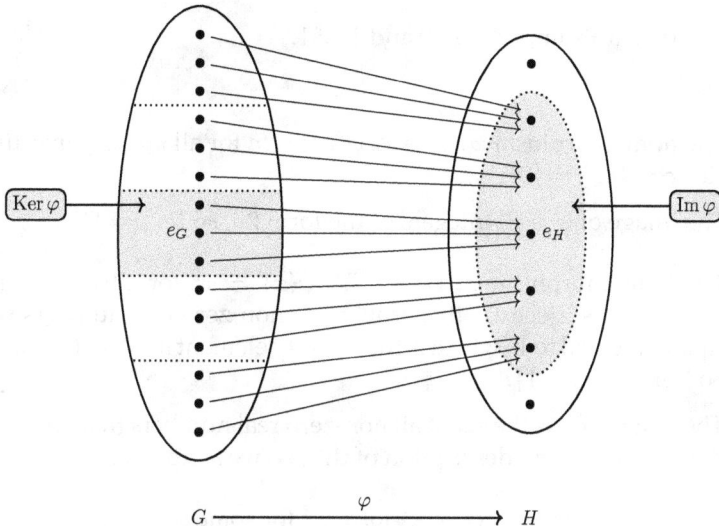

Very shortly we will come to examples, but first we will record some basic facts about kernels and images that will be very helpful in working out the examples.

Proposition 1.2.63. *Let* $\varphi\colon G \to H$ *be a group homomorphism.*

   (i) *The kernel of* φ, $\operatorname{Ker} \varphi$, *is a subgroup of* G.

  (ii) *The image of* φ, $\operatorname{Im} \varphi$, *is a subgroup of* H.

 (iii) *The homomorphism* φ *is injective if and only if* $\operatorname{Ker} \varphi = \{e_G\}$.

 (iv) *The homomorphism* φ *is surjective if and only if* $\operatorname{Im} \varphi = H$.

  (v) *The homomorphism* φ *is an isomorphism if and only if* $\operatorname{Ker} \varphi = \{e_G\}$ *and* $\operatorname{Im} \varphi = H$.

Remark 1.2.64. Let $\varphi\colon G \to H$ be an *injective* homomorphism. Then we can consider the function $\psi = \varphi|^{\operatorname{Im} \varphi}\colon G \to \operatorname{Im} \varphi$, the codomain restriction to $\operatorname{Im} \varphi$ of φ. (This is the function that takes the same values as φ but where we "throw away" any elements of H not in the image of φ, and so just take $\operatorname{Im} \varphi$ as the codomain.) Since $\operatorname{Ker} \psi = \operatorname{Ker} \varphi = \{e_G\}$ and $\operatorname{Im} \psi = \operatorname{Im} \varphi$, ψ is both injective and surjective. Hence ψ is an isomorphism of G with $\operatorname{Im} \varphi$.

Since $\operatorname{Im} \varphi$ is a subgroup of H, in this situation we say that G is isomorphic to a subgroup of H.

That is, to show that a group G is isomorphic to a subgroup of a group H, we show that there exists an injective homomorphism from G to H. If that homomorphism is also surjective, we have that G is isomorphic to H itself.

Let us revisit Examples 1.2.50 and 1.2.51.

Examples 1.2.65.

 (a) The homomorphism $\varphi\colon \mathbb{Z} \to \mathbb{Z}$, $\varphi(n) = 2n$ for all $n \in \mathbb{Z}$ is injective, so $\operatorname{Ker} \varphi = \{e_{\mathbb{Z}}\} = \{0\}$.

    The image of φ is all integers of the form $2n$, so $\operatorname{Im} \varphi = 2\mathbb{Z}$.

 (b) The homomorphism $\varphi\colon \mathbb{R}^* \to \mathbb{R}^*$, $\varphi(x) = x^2$ for all $x \in \mathbb{R}^*$ is not injective or surjective. Its kernel is all non-zero real numbers whose square is equal to 1 (which is the identity element in $\mathbb{R}^* = (\mathbb{R}\backslash\{0\}, \times)$). So $\operatorname{Ker} \varphi = \{1, -1\}$.

    The image of φ is the set of all non-zero real numbers that are squares; there is no "nicer" description of this, so we simply have

$$\operatorname{Im} \varphi = \{y \in \mathbb{R}^* \mid y = x^2 \text{ for some } x \in \mathbb{R}^*\}.$$

(c) The homomorphism $\varphi\colon \mathbb{R} \to \mathbb{C}^*$, $\varphi(x) = e^{2\pi i x}$ for all $x \in \mathbb{R}$ has as kernel all integers, $\operatorname{Ker}\varphi = \mathbb{Z}$.

As we saw before, the image of φ is

$$\operatorname{Im}\varphi = \{z \in \mathbb{C} \mid \|z\| = 1\} \subsetneq \mathbb{C}^*.$$

(d) The homomorphism $\varphi\colon S_n \to \operatorname{GL}_n(\mathbb{R})$, $\varphi(\sigma) = A_\sigma$ is injective, so $\operatorname{Ker}\varphi = \{e_{S_n}\} = \{\iota\}$. Again the image of φ has no particularly nice description: it is just

$$\operatorname{Im}\varphi = \{A \in \operatorname{GL}_n(\mathbb{R}) \mid A = A_\sigma \text{ for some } \sigma \in S_n\}.$$

(e) The homomorphism $\Sigma\colon S_n \to \{1, -1\}$, $\Sigma(\pi) = \operatorname{sign}(\pi)$ for all $\pi \in S_n$ is not injective for $n \geq 3$. Its kernel is

$$\operatorname{Ker}\Sigma = \{\sigma \in S_n \mid \operatorname{sign}(\sigma) = 1\}$$

since 1 is the identity element in $(\{1, -1\}, \times)$. By definition, these are the even permutations, A_n, so $\operatorname{Ker}\Sigma = A_n$. If $n \geq 2$, since there exist both even and odd permutations, Σ is surjective and $\operatorname{Im}\Sigma = \{1, -1\}$.

In the picture above, we saw that given a homomorphism $\varphi\colon G \to H$, we can split G up into pieces labelled by the element of H that the elements of that piece map to under φ. The next result tells us that this partition of G, coming from the equivalence relation $\mathcal{R}_\varphi$ defined by $g_1 \,\mathcal{R}_\varphi\, g_2$ if and only if $\varphi(g_1) = \varphi(g_2)$, is a particularly nice partition: it is the same one as we obtain from $\mathcal{R}_{\operatorname{Ker}\varphi}$ as described in Theorem 1.2.35.

Proposition 1.2.66. Let $\varphi\colon G \to H$ be a group homomorphism and let $K = \operatorname{Ker}\varphi$. Then for all $g \in G$,

$$gK = \{l \in G \mid g \,\mathcal{R}_\varphi\, l\} = \{l \in G \mid \varphi(g) = \varphi(l)\}.$$

Hence the function $\psi\colon L_K \to \operatorname{Im}\varphi$, $\psi(gK) = \varphi(g)$ is a bijection, where $L_K = \{gK \mid g \in G\}$ is the set of left cosets of $K = \operatorname{Ker}\varphi$.

Since the index $|G : \operatorname{Ker}\varphi|$ is defined to be the cardinality of the set of left cosets of $\operatorname{Ker}\varphi$, as an immediate corollary of this result and Lagrange's theorem we have:

Corollary 1.2.67. Let $\varphi\colon G \to H$ be a group homomorphism. We have that $|G : \operatorname{Ker}\varphi| = |\operatorname{Im}\varphi|$ and hence $|G| = |\operatorname{Ker}\varphi||\operatorname{Im}\varphi|$.

This is the analogous result for groups to the Dimension Theorem for vector spaces, which asserts that for a linear transformation $T\colon V \to W$ we have $\dim V = \dim \operatorname{Ker}T + \dim \operatorname{Im}T$.

In fact, both of these are "numerical shadows" of stronger statements: we will shortly introduce quotient groups and see the stronger version for groups.

Since $\text{Ker } \varphi \leq G$ and $\text{Im } \varphi \leq H$, we already knew that $|\text{Ker } \varphi|$ divides $|G|$ and $|\text{Im } \varphi|$ divides $|H|$, by Lagrange's theorem. However this corollary tells us that $|\text{Im } \varphi|$ also divides $|G|$, which can be useful to know, in applications such as the following.

Example 1.2.68. Let G be a group of order 16 and H a group of order 9. Then the only homomorphism $\varphi \colon G \to H$ is the trivial homomorphism with $\varphi(g) = e_H$ for all $g \in G$. For if $\varphi \colon G \to H$ is a homomorphism, $|\text{Im } \varphi|$ must divide $|G|$ and $|H|$, but $\text{hcf}(16, 9) = 1$ so $|\text{Im } \varphi| = 1$ and so $\text{Im } \varphi = \{e_H\}$.

The kernel of a group homomorphism has a stronger property than just being a subgroup, namely *normality*, expressed as follows in terms of cosets.

Definition 1.2.69. Let N be a subgroup of a group G. We say that N is a normal subgroup if $gN = Ng$ for all $g \in G$. We write $N \trianglelefteq G$, or $N \triangleleft G$ if N is a proper normal subgroup.

Note in particular, as it will be relevant for the other algebraic structures we introduce below, that every subgroup of an Abelian group is normal. As another example, the centre of a group G,

$$Z(G) = \{z \in G \mid zg = gz \text{ for all } g \in G\},$$

is an Abelian subgroup of G and furthermore $Z(G)$ is a normal subgroup of G.

Importantly, we have the following.

Proposition 1.2.70. *Let* $\varphi \colon G \to H$ *be a group homomorphism. Then the kernel of* φ, $\text{Ker } \varphi$, *is a normal subgroup of* G.

1.2.6   Cayley's theorem

Sometimes one needs to calculate in particular explicit examples. This might be done by hand or, more commonly now, by a computer package. But in order to do so, we need a "nice" way to represent the elements of our group, to store them and to calculate the result of the binary operation on them. In this section, we will see two theoretical results that can be used to provide such representations. The first is called Cayley's theorem.

Definition 1.2.71. Let X be a set. The symmetric group on X is defined to be the group $S_X = (\text{Bij}(X), \circ)$ of bijections between X and itself.

Theorem 1.2.72 (Cayley's theorem). *Let* G *be a group. There is an injective homomorphism* $\lambda \colon G \to S_G$.

Corollary 1.2.73. *Every group G is isomorphic to a subgroup of S_G.*

In particular, every finite group G of order n is isomorphic to a subgroup of the symmetric group of degree n, S_n.

Corollary 1.2.74. *Let $n \in \mathbb{N}$. There are finitely many non-isomorphic groups of order n.*

The word "non-isomorphic" here is very important: there are infinitely many groups with one element, for example, but they are all isomorphic.[3]

We can also join two previous results together: Cayley's theorem and the homomorphism from the symmetric group to the general linear group given by taking permutation matrices. This gives us, for free, a matrix version of Cayley's theorem.

Theorem 1.2.75. *Let G be a finite group. There is an injective homomorphism $\mu \colon G \to \mathrm{GL}_{|G|}(\mathbb{R})$.*

1.2.7 Quotient groups

Shortly we will see that the above result about kernels of homomorphisms being normal subgroups is only half of the story: in fact, every normal subgroup is the kernel of some homomorphism. To explain this will take some more preparation, however. Firstly, we will, as promised, see that if we have a normal subgroup N of a group G, we can put an algebraic structure on the set of cosets of N in G. Specifically, the set of cosets inherits a group structure from G and this new group is called the quotient group.

Although the construction may seem abstract, we will soon see that a well-known group is most properly understood as a quotient, namely the integers modulo n.

The ingredients for the construction are:

- a group $(G, *)$ (to try to avoid confusion, we will revert to using an explicit symbol to denote the binary operation in G);

- a normal subgroup N of G; and

- the set of (left) cosets $L_N = \{gN \mid g \in G\}$ of N in G.

Proposition 1.2.76. *Let G be a group and N a normal subgroup of G. Then $\bullet \colon L_N \times L_N \to L_N$ given by $gN \bullet hN = (g * h)N$ defines a binary operation on L_N.*

Definition 1.2.77. *Let $(G, *)$ be a group and let N be a normal subgroup of G. The quotient group of G by N, denoted G/N, is the group $(L_N, \bullet)$ where L_N is the set $\{gN \mid g \in G\}$ of left cosets of N in G and $\bullet$ is the binary operation $gN \bullet hN = (g * h)N$.*

---

[3]For each $x \in \mathbb{R}$, let $E_x = (\{x\}, *)$ be the group with binary operation $x * x = x$ (note that $*$ is neither $+$ or $\times$!). Then if $x \neq y$, $E_x \neq E_y$ but $E_x \cong E_y$ for all $x, y \in \mathbb{R}$ via $x \mapsto y$.

Proposition 1.2.78. *The binary operation* $gN \bullet hN = (g * h)N$ *defines a group structure on* $L_N = \{gN \mid g \in G\}$.

From now on, we will use the notation G/N in preference to $(L_N, \bullet)$, so that when we write G/N it is understood that G/N as a set is the set of left cosets of a normal subgroup N of G and that the binary operation is $\bullet$.

Corollary 1.2.79. *We have* $|G/N| = |G|/|N|$.

Definition 1.2.80. Let $n \in \mathbb{Z}$. The group of integers modulo n is defined to be the quotient group $(\mathbb{Z}/n\mathbb{Z}, +_n)$. We also often denote this group by $(\mathbb{Z}_n, +_n)$.

Recall that $\mathbb{Z}_n \cong C_n$ is cyclic of order n, with one choice of generator being $1 + n\mathbb{Z}$.

1.2.8 Presentations of groups

More than once we have implicitly used the notion of a *presentation* of a group, for example when we described the dihedral groups and most recently the cyclic groups. We can now say a little more precisely what this means, although we will not check all the details.

Giving a presentation for a group G, written $\langle X \mid \mathcal{R} \rangle$, is the claim that G is isomorphic to a *quotient* of the *free group* generated by X, $\mathcal{F}(X) = \langle X \rangle$.

The free group on a set is analogous to forming monomials in a set of variables: the elements of the free group are *words* in the set X, which simply means finite expressions of the form $x_1^{\pm 1} x_2^{\pm 1} \cdots x_r^{\pm 1}$ with $x_i \in X$. This becomes a group by the operation of *concatenation* (sticking two such expressions together), where we understand that the expression of length 0 is the identity e and $x_i x_i^{-1} = e = x_i^{-1} x_i$.

The particular quotient we take to form $\langle X \mid \mathcal{R} \rangle$ is the one where we take $\mathcal{R}$ to be a set of elements of $\mathcal{F}(X)$, form the subgroup $\langle \mathcal{R} \rangle$ they generate and then take the *normal closure* of this, i.e. we find the smallest normal subgroup $\overline{\langle \mathcal{R} \rangle}$ of $\mathcal{F}(X)$ containing $\langle \mathcal{R} \rangle$. Then $\langle X \mid \mathcal{R} \rangle \stackrel{\text{def}}{=} \mathcal{F}(X)/\overline{\langle \mathcal{R} \rangle}$.

Elements of X are called generators of $\langle X \mid \mathcal{R} \rangle$ and elements of $\mathcal{R}$ are called *relators*. In practice, we often give elements of $\mathcal{R}$ as *relations* rather than relators. Defining by way of an example, if we want a relation such as $b^{-1}ab = a^{-1}$ ("imposing a relation" meaning that in the group we want the two sides to be equal) then formally we should make this into the relator $b^{-1}aba$ and put this element in $\mathcal{R}$. Then in the quotient group $b^{-1}aba$ will become equal to the identity element of the quotient, so $b^{-1}aba = e$, i.e. $b^{-1}ab = a^{-1}$. (We are being sloppy and using the same letters for the elements a, b of $\mathcal{F}(X)$ and their images $a + \overline{\langle \mathcal{R} \rangle}, b + \overline{\langle \mathcal{R} \rangle}$ in the quotient, but this is common practice.) So rather than $C_n = \langle g \mid g^n = e \rangle$ we should write $\langle g \mid g^n \rangle$, but the former is easier to understand and we do that instead.

1.2.9   The First Isomorphism Theorem for Groups

First we will record the aforementioned claim that every normal subgroup is the kernel of a homomorphism. Then in the First Isomorphism Theorem we will relate kernels, images and quotients.

Lemma 1.2.81. *Let G be a group and N a normal subgroup of G. The function $\pi\colon G \to G/N$ defined by $\pi(g) = gN$ is a group homomorphism.*

Definition 1.2.82. Let G be a group and N a normal subgroup of G. The homomorphism $\pi\colon G \to G/N$ defined by $\pi(g) = gN$ is called the quotient homomorphism associated to G and N.

Proposition 1.2.83. *Let G be a group and N a normal subgroup of G. The quotient homomorphism $\pi\colon G \to G/N$ is surjective and its kernel is $\operatorname{Ker}\pi = N$.*

Corollary 1.2.84. *Let G be a group and N a normal subgroup of G. Then N is the kernel of a group homomorphism with domain G, namely the quotient homomorphism $\pi\colon G \to G/N$.*

Theorem 1.2.85 (Universal property of the quotient group)**.** Let G and H be *groups. Let N be a normal subgroup of G and let $\pi\colon G \to G/N$ be the associated quotient homomorphism.*

Then for every homomorphism $\varphi\colon G \to H$ such that $N \subseteq \operatorname{Ker}\varphi$, there exists a unique homomorphism $\bar\varphi\colon G/N \to H$ such that $\varphi = \bar\varphi \circ \pi$.

Furthermore, $\operatorname{Ker}\bar\varphi = (\operatorname{Ker}\varphi)/N$ and $\operatorname{Im}\bar\varphi = \operatorname{Im}\varphi$.

We can illustrate the maps in this theorem via the following diagram, known as a "commuting diagram" (the "commuting" refers to the fact that following along the arrows in either possible way gives the same result).

$$
\begin{array}{ccc}
G & \overset{\varphi}{\longrightarrow} & H \\
{\scriptstyle \pi}\big\downarrow & \nearrow & \\
G/N & {\scriptstyle \exists!\bar\varphi} &
\end{array}
$$

A consequence of this theorem is that there is a one-to-one correspondence between homomorphisms $\varphi\colon G \to H$ with $N \subseteq \operatorname{Ker}\varphi$ and homomorphisms $\bar\varphi\colon G/N \to H$, the correspondence being given by composing with π. Since quotient groups can be complicated to understand, this tells us that to find homomorphisms whose domain is G/N, we can instead look for homomorphisms whose domain is G and just check if N is a subgroup of the kernel of these. (Depending on the situation, the converse direction can be very helpful too, of course.)

The above theorem has done much of the heavy lifting to enable us to deduce in a nice, clean fashion the First Isomorphism Theorem.[4]

---

[4]Arguably, it should be called the "First Isomorphism Corollary" since it follows essentially immediately from the previous theorem, but often theorems are called theorems for their significance rather than their outright difficulty.

Theorem 1.2.86 (First Isomorphism Theorem for Groups). *Let G and H be groups and let $\varphi\colon G \to H$ be a group homomorphism. Then*

$$G/\operatorname{Ker}\varphi \cong \operatorname{Im}\varphi.$$

Challenge 5. SageMath returns True when evaluating the following code. Find a homomorphism that explains this.

```
1  A4 = AlternatingGroup(4)
2  r1 = A4("(1,2)(3,4)")
3  r2 = A4("(1,3)(2,4)")
4  r3 = A4("(1,4)(2,3)")
5  K = A4.subgroup([r1, r2, r3])
6  I = A4.quotient(K)
7  C3 = CyclicPermutationGroup(3)
8  I.is_isomorphic(C3)
```

1.3   Rings and fields

Now, we recap the foundational theory of rings, with some extra observations for (the special case of) fields. We can "bootstrap" much of what we need from group theory, as above, so we will still be relatively brief. We first recall the definition of a ring as we gave it before.

Definition. A ring is an algebraic structure $(R, +, \times)$ such that $(R, +)$ is an Abelian group, $\times$ is associative and we have the distributive laws

$$a \times (b + c) = (a \times b) + (a \times c) \qquad \text{and} \qquad (a + b) \times c = (a \times c) + (b \times c)$$

for all $a, b, c \in R$. The identity element for the group $(R, +)$ will be denoted 0_R. The inverse of a with respect to $+$ will be denoted by $-a$.

Remark 1.3.1. The distributive laws can seem a little like they appear out of thin air. However, if one stares at them for a while, one sees that in fact they say that left and right multiplication are group homomorphisms. More precisely, given $a \in R$, the function $\mu_a^L\colon R \to R$ given by $\mu_a^L(b) = a \times b$ is a homomorphism from the Abelian group $R = (R, +)$ to itself, and similarly for $\mu_c^R\colon R \to R$, $\mu_c^R(b) = b \times c$. From this point of view, the conditions above are not so unnatural.

The following discussion on power series rings and polynomial rings gives us some important classes of rings to use as examples and in subsequent exercises. Although we will not take any significant steps in this direction, we comment that these also play fundamental roles in algebraic geometry.

Definition 1.3.2. Let R be a ring. The set $S(R)$ of ($\mathbb{N}$-indexed) sequences of elements of R, with the operations $+_{S(R)}$ and $\times_{S(R)}$ defined by

$$(\underline{a} +_{S(R)} \underline{b})_n = a_n +_R b_n$$

and

$$(\underline{a} \times_{S(R)} \underline{b})_n = \sum_{i+j=n} a_i \times_R b_j$$

is a ring, called the ring of formal power series (in one variable) over R. We denote this ring by $R[[x]]$ and the ring R is called the base ring of $R[[x]]$.

Remark 1.3.3. If R has a multiplicative identity 1_R (see Definition 1.3.10) then the "variable" x may be identified with the sequence

$$x_n = \begin{cases} 1_R & \text{for } n = 1 \\ 0_R & \text{otherwise.} \end{cases}$$

This covers most familiar situations but has the perhaps surprising implication that if R does *not* have a multiplicative identity, then $x = 0_R + 1_R x + 0_R x^2 + 0_R x^3 + \cdots \notin R[[x]]$.

Now that we have the ring of formal power series, it is straightforward to define the ring of polynomials. Polynomials are just "special" power series, namely they are the power series that are "eventually zero", i.e. after some point, every element of the sequence is 0_R. Another way to say this is that only finitely many elements of the corresponding sequence are non-zero. We say that a sequence $\underline{a}$ is finitely supported if it has this property: $\underline{a}$ is finitely supported if $|\{i \mid a_i \neq 0_R\}|$ is finite.

Definition 1.3.4. Let R be a ring. The set $S_{\text{fs}}(R)$ of finitely supported ($\mathbb{N}$-indexed) sequences of elements of R, with the operations $=_{S_{\text{fs}}(R)}$ and $\times_{S_{\text{fs}}(R)}$ defined by

$$(\underline{a} +_{S_{\text{fs}}(R)} \underline{b})_n = a_n +_R b_n$$

and

$$(\underline{a} \times_{S_{\text{fs}}(R)} \underline{b})_n = \sum_{i+j=n} a_i \times_R b_j$$

is a ring, called the ring of polynomials (in one variable) over R. We denote this ring by $R[x]$ and the ring R is called the base ring of $R[x]$.

Indeed, $S_{\text{fs}}(R) = R[x]$ is a *subring* of $S(R) = R[[x]]$ (see Section 1.3.2).

SageMath can handle polynomial rings over $\mathbb{Z}$ (ZZ in SageMath), $\mathbb{Q}$ (QQ) and $\mathbb{R}$ (RR), as follows. Note that we have to tell SageMath what we want our variable to be called. The code

```
1  R.<x> = PolynomialRing(QQ)
2  S.<y> = PolynomialRing(RR)
3  p = 3*x^2+x+1
4  q = sqrt(3)*x+2
5  r = sqrt(3)*y+2
6  [p in R, r in S, p in S, q in R]
```

constructs two polynomial rings, $R = \mathbb{Q}[x]$ and $S = \mathbb{R}[y]$, and three polynomials p, q, r such that $p \in R$ and $r \in S$ but $p \notin S$ (since p is a polynomial in the variable x, not y) and $q \notin R$ (since $\sqrt{3} \notin \mathbb{Q}$).

Remark 1.3.5. As in Remark 1.3.3, if R has a multiplicative identity 1_R then the "variable" x may be identified with the (finitely supported) sequence

$$x_n = \begin{cases} 1_R & \text{for } n = 1 \\ 0_R & \text{otherwise.} \end{cases}$$

Again this leads to the counter-intuitive observation that if R does *not* have a multiplicative identity, then $x \notin R[x]$, i.e. x is not a polynomial—the point being that it is not a polynomial *with coefficients in R*. For a concrete example, consider the ring $2\mathbb{Z}[x]$ of polynomials in one variable x with even integer coefficients: $x \notin 2\mathbb{Z}[x]$.

Definition 1.3.6. Let R be a ring and $R[x]$ the polynomial ring in one variable over R. Let $p = (p_n)$ be a non-zero element of $R[x]$, i.e. there exists i such that $p_i \neq 0_R$.

The degree of p is defined to be

$$\deg p = \max\{k \mid p_k \neq 0_R\}.$$

For $m = \max\{k \mid p_k \neq 0_R\}$ we say that p_m is the leading coefficient of p and $p_m x^m$ the leading term of p. If $m = 0$, we say that p is a constant polynomial.

Continuing after the above code, entering

```
7  p.degree()
```

returns 2, as expected.

We can also ask SageMath to factor and find roots (with multiplicities), if these exist in the base ring:

```
1   R.<x> = PolynomialRing(RR)
2   p = x^2+3*x+2
3   p.factor()
4   p.roots()
5   q = x^2-2*x+1
6   q.factor()
7   q.roots()
8   S.<y> = PolynomialRing(CC)
9   r = y^2+1
10  r.factor()
11  r.roots()
```

Here, CC means the complex numbers, $\mathbb{C}$, and I (in the output) represents i.

Proposition 1.3.7. *Let R be a ring such that for any two elements $a, b \in R$ such that $a, b \neq 0_R$, we have $ab \neq 0_R$. Let $R[x]$ be the ring of polynomials in one variable over R and let $p, q \in R[x]$. Then*

(i) $\deg(p + q) \leq \max\{\deg p, \deg q\}$, *and*

(ii) *if $p, q \neq 0_{R[x]}$ then $pq \neq 0_{R[x]}$ and*

$$\deg pq = \deg p + \deg q.$$

Proposition 1.3.8 (The division algorithm for polynomials). *Let F be a field and let $f, g \in F[x]$, with $g \neq 0$. Then there exist unique polynomials $q, r \in F[x]$ such that $f = qg + r$ and either $r = 0$ or $\deg r < \deg g$.*

The polynomial q is called the quotient, and r the remainder, of f divided by g.

1.3.1 Basic properties of rings

Proposition 1.3.9. *Let $(R, +, \times)$ be a ring and 0_R the identity element for $+$. Recall that we denote the inverse of a with respect to $+$ by $-a$.*

(i) (additive cancellation) *For all $a, b, c \in R$, if $a + b = a + c$ then $b = c$.*

(ii) *For all $a \in R$, $0_R \times a = 0_R = a \times 0_R$.*

(iii) *For all $a, b \in R$, $a \times (-b) = (-a) \times b = -(a \times b)$ and $(-a) \times (-b) = a \times b$.*

As above, we take particular care over the existence of a multiplicative identity.

Definition 1.3.10. Let $(R, +, \times)$ be a ring. An element $1_R \in R$ is called a multiplicative identity if 1_R is an identity for the binary operation $\times$, that is, for all $a \in R$,

$$a \times 1_R = a = 1_R \times a.$$

A ring having a multiplicative identity is said to be *unital*.

Notice that a multiplicative identity is unique if it exists: an identity for a binary operation is always unique.

Examples 1.3.11.

(a) The trivial ring $(\{e\}, +, \times)$ with $e + e = e$ and $e \times e = e$ has multiplicative identity e, as we see from the definition of $\times$. One can show that the only ring R with a multiplicative identity 1_R in which $1_R = 0_R$ is the trivial ring.

(b) The ring $(\mathbb{Z}, +, \times)$, with the usual addition and multiplication of integers, has multiplicative identity 1; similarly for $\mathbb{Q}$, $\mathbb{R}$ and $\mathbb{C}$, with their usual operations.

(c) The integers modulo n, $\mathbb{Z}_n$, with addition and multiplication modulo n has a multiplicative identity $\widehat{1} = 1 + n\mathbb{Z}$. Later we will see why this should not be a surprise.

(d) The matrix rings $\mathrm{M}_n(\mathbb{Z})$, $\mathrm{M}_n(\mathbb{Q})$, $\mathrm{M}_n(\mathbb{R})$ and $\mathrm{M}_n(\mathbb{C})$ all have multiplicative identities, namely the $n \times n$ identity matrix I_n,

$$(I_n)_{ij} = \begin{cases} 1 & \text{if } i = j \\ 0 & \text{otherwise.} \end{cases}$$

The matrix ring $\mathrm{M}_n(R)$, for R an arbitrary ring, need not have a multiplicative identity; shortly we will see precisely when it does.

(e) The ring of functions $\mathcal{F}(Y, \mathbb{R})$ (respectively $\mathcal{F}(Y, \mathbb{C})$) (where Y is any set) has a multiplicative identity, the constant function $\mathbb{1} : Y \to \mathbb{R}$ (respectively $\mathbb{1} : Y \to \mathbb{C}$) defined by $\mathbb{1}(y) = 1$ for all $y \in Y$.

(f) The polynomial rings $\mathbb{Z}[x]$, $\mathbb{Q}[x]$, $\mathbb{R}[x]$ and $\mathbb{C}[x]$ have multiplicative identities, namely the constant polynomial $1 = 1 + 0x + 0x^2 + \cdots$. The polynomial ring $R[x]$, for R an arbitrary ring, need not have a multiplicative identity; again we shall shortly see precisely when it does.

Just as in the case of groups, rings with a commutative multiplication operation are particularly special. Remember that a ring $(R, +, \times)$ is by definition an *Abelian* group with respect to $+$, so $a + b = b + a$ for all a, b always. The issue at hand for rings is whether or not the multiplication operation $\times$ is commutative.

Definition 1.3.12. Let $(R, +, \times)$ be a ring. We say that R is a commutative ring if the multiplication operation $\times$ is commutative. That is, $a \times b = b \times a$ for all $a, b \in R$.

Be aware that it is common practice to say "non-commutative ring" when what is actually meant is "ring" (that is, the ring in question might be commutative or it might not be, we don't know and/or don't care, but in particular we won't assume it is commutative). In this usage, "non-commutative" is not the negation of "commutative": "non-commutative" and "not commutative" are then unfortunately not synonyms.

Note that the negation of $(\forall a, b \in R)(a \times b = b \times a)$ is

$$\neg\,((\forall a, b \in R)(a \times b = b \times a)) = (\exists\, a, b \in R)(a \times b \neq b \times a),$$

so that to show a ring R is *not* commutative we must find two specific elements of R whose products $a \times b$ and $b \times a$ are not equal. (We must have $a \neq b$, since $a \times a = a \times a$, of course.)

We have already seen a number of examples of commutative and non-commutative rings.

Examples 1.3.13.

(a) $\mathbb{Z}, \mathbb{Q}, \mathbb{R}$ and $\mathbb{C}$ are commutative (unital) rings.

(b) The integers modulo n, $\mathbb{Z}_n$, are a commutative ring.

(c) Matrix rings are usually *non-commutative*.

(d) The rings of functions $\mathcal{F}(Y, \mathbb{R})$ and $\mathcal{F}(Y, \mathbb{C})$ are commutative: multiplication is defined pointwise so if $f, g \in \mathcal{F}(Y, \mathbb{R})$,

$$(fg)(y) = f(y)g(y) = g(y)f(y) = (gf)(y)$$

for all $y \in Y$; similarly for $f, g \in \mathcal{F}(Y, \mathbb{C})$. A little careful thought shows that the key issue here for $\mathcal{F}(Y, R)$ with R an arbitrary ring is whether or not R is commutative.

(e) Polynomial rings $R[x]$ can be commutative or not, depending on R.

(f) The ring of even integers $2\mathbb{Z}$ is commutative, because its multiplication is that of $\mathbb{Z}$, restricted to the even integers, and multiplication of integers is commutative.

Lemma 1.3.14. *Let R be a ring such that there exist $r, s \in R$ with $rs \neq 0_R$. Then the matrix ring $M_n(R)$ is commutative if and only if $n = 1$ and R is commutative. That is, $M_n(R)$ is non-commutative if $n \geq 2$, and $M_1(R)$ is non-commutative if and only if R is.*

Lemma 1.3.15. *Let R be a ring. Then the polynomial ring $R[x]$ is commutative if and only if R is.*

1.3.2 Subrings

Just as with groups and subgroups, it is natural to study subsets of rings that are themselves rings. Indeed, in a few places above it would have been natural to say "subring", so it is not before time that we give the definition.

Definition 1.3.16. Let $(R, +, \times)$ be a ring. A subset S of R is said to be a subring if $(S, +, \times)$ is a ring. If this is the case, we write $S \leqslant R$.

Proposition 1.3.17 (Subring test)**.** *Let S be a subset of a ring $(R, +, \times)$ and denote the additive identity element of R by 0_R. Then S is a subring of R if and only if the following two conditions are satisfied:*

(SR1) *S is a subgroup of $(R, +)$;*

(SR2) *for all $s_1, s_2 \in S$, we have $s_1 \times s_2 \in S$.*

Lemma 1.3.18. *Every subring of a commutative ring is commutative.*

Problem 6. Let R and S be rings, and let T denote their Cartesian product, that is,

$$T = R \times S = \{(a, b) \mid a \in R,\ b \in S\}.$$

Equip T with the coordinate-wise defined addition and multiplication operations

$$(a, b) +_T (c, d) = (a +_R c, b +_S d)$$

and

$$(a, b) \times_T (c, d) = (a \times_R c, b \times_S d)$$

for $a, c \in R$ and $b, d \in S$.

(a) Show that T is a ring with respect to these operations.

(b) Show that T is commutative if and only if R and S are commutative.

1.3.3   Homomorphisms

Earlier, in Section 1.2.4, we saw that the correct way to relate two groups $(G, *)$ and $(H, \circ)$ is to define a group homomorphism, this being a function $\varphi \colon G \to H$ such that $\varphi(g_1 * g_2) = \varphi(g_1) \circ \varphi(g_2)$ for all $g_1, g_2 \in G$. A ring $(R, +, \times)$ is by definition an Abelian group $(R, +)$ with a compatible multiplication $\times$, so a ring homomorphism must be a group homomorphism that also preserves the multiplication operations.

Definition 1.3.19. Let $(R, +_R, \times_R)$ and $(S, +_S, \times_S)$ be rings. A function $\varphi \colon R \to S$ is called a ring homomorphism if φ is a group homomorphism and preserves the multiplication operations. Explicitly,

(H1) $\varphi(r_1 +_R r_2) = \varphi(r_1) +_S \varphi(r_2)$ and

(H2) $\varphi(r_1 \times_R r_2) = \varphi(r_1) \times_S \varphi(r_2)$,

for all $r_1, r_2 \in R$.

Let us again list some elementary properties of ring homomorphisms, the first two of which are immediate from a ring homomorphism being in particular a group homomorphism, so are restatements of parts of Proposition 1.2.48. The third is easily proved by induction.

Proposition 1.3.20. *Let* $\varphi\colon R \to S$ *be a ring homomorphism. Then*

(i) $\varphi(0_R) = 0_S$;

(ii) $\varphi(-r) = -\varphi(r)$ *for all* $r \in R$;

(iii) $\varphi(r^n) = \varphi(r)^n$ *for all* $r \in R, n \in \mathbb{N}$.

Proposition 1.3.21. *Let* $\varphi\colon R \to S$ *and* $\sigma\colon S \to T$ *be ring homomorphisms. Then* $\sigma \circ \varphi\colon R \to T$ *is a ring homomorphism.*

Example 1.3.22. Next we give the ring analogue of the trivial group homomorphism defined in Exercise 4.

Let R and S be rings and let $\varphi\colon R \to S$ be the function defined by $\varphi(a) = 0_S$ for all $a \in R$. Then φ is a ring homomorphism because for all $a, b \in R$, we have

(H1) $\varphi(a + b) = 0_S = 0_S + 0_S = \varphi(a) + \varphi(b)$ and

(H2) $\varphi(ab) = 0_S = 0_S \cdot 0_S = \varphi(a)\varphi(b)$.

We call φ the *zero homomorphism*; all other homomorphisms are said to be *non-zero*.

Example 1.3.23. Recall that for any set B and any subset $A \subseteq B$, there is an (injective) function $\iota\colon A \to B$ defined by $\iota(a) = a$ for all $a \in A$. This is called the inclusion map, as it precisely encodes A being "included" in B, as a subset.

Let S be a subring of a ring R, so that in particular $S \subseteq R$, and let $\iota\colon S \to R$ be the inclusion map. Then ι is a ring homomorphism, as is easily checked.

Definition 1.3.24. A ring homomorphism $\varphi\colon R \to S$ is called a ring isomorphism if there exists a ring homomorphism $\psi\colon S \to R$ such that $\psi \circ \varphi = \mathrm{id}_R$ and $\varphi \circ \psi = \mathrm{id}_S$.

We say that two rings R and S are isomorphic if and only if there exists a ring isomorphism $\varphi\colon R \to S$. If R and S are isomorphic, we often write $R \cong S$.

Again, the identity function $\mathrm{id}_R\colon R \to R$, $\mathrm{id}_R(r) = r$ is a ring homomorphism and is invertible, so is a ring isomorphism.

The following lemma follows immediately from Lemma 1.2.54, since we simply add the property (H2) to the conditions in the corresponding statement for group homomorphisms.

Lemma 1.3.25. *Let R and S be rings. Then the following are are equivalent:*

(i) *R and S are isomorphic, i.e. there exist ring homomorphisms $\varphi\colon R \to S$ and $\psi\colon S \to R$ such that $\psi \circ \varphi = \mathrm{id}_R$ and $\varphi \circ \psi = \mathrm{id}_S$; and*

(ii) *there exists a bijective ring homomorphism $\varphi\colon R \to S$.*

Definition 1.3.26. Let R and S be rings. If S contains a subring T such that T is isomorphic to R, then we say that S contains a subring isomorphic to R.

In practice we use the following criterion to decide whether or not a ring contains a subring isomorphic to some other ring.

Proposition 1.3.27. *Let R and S be rings. Then S contains a subring isomorphic to R if and only if there is an injective ring homomorphism from R to S.*

If R and S are unital rings, we may or may not have that a ring homomorphism sends 1_R to 1_S. If we want to be sure of this, we need to ask for it explicitly:

Definition 1.3.28. Let R and S be unital rings. We say that a ring homomorphism $\varphi\colon R \to S$ is *unital* if $\varphi(1_R) = 1_S$.

1.3.4   Kernels and images

We may take the same definitions (as sets) for the kernel and the image of a ring homomorphism, by simply considering it as a group homomorphism, "forgetting" its extra property of being compatibile with multiplication. However, we use precisely the extra multiplicativity to show that the kernel and image have more structure than just being subgroups.

Definition 1.3.29. Let $\varphi\colon R \to S$ be a ring homomorphism. The kernel of φ is defined to be the subset of R given by

$$\mathrm{Ker}\,\varphi = \{r \in R \mid \varphi(r) = 0_S\}.$$

Definition 1.3.30. Let $\varphi\colon R \to S$ be a ring homomorphism. The image of φ is defined to be the subset of S given by

$$\mathrm{Im}\,\varphi = \{s \in S \mid (\exists r \in R)(\varphi(r) = s)\}.$$

Proposition 1.3.31. *Let $\varphi\colon R \to S$ be a ring homomorphism.*

(i) *The kernel of φ, $\mathrm{Ker}\,\varphi$, is a subring of R.*

(ii) *The image of φ, $\mathrm{Im}\,\varphi$, is a subring of S.*

(iii) *The homomorphism φ is injective if and only if $\mathrm{Ker}\,\varphi = \{0_R\}$.*

(iv) *The homomorphism φ is surjective if and only if $\operatorname{Im} \varphi = S$.*

(v) *The homomorphism φ is an isomorphism if and only if $\operatorname{Ker} \varphi = \{0_R\}$ and $\operatorname{Im} \varphi = S$.*

Proposition 1.3.32. *Let R be a ring. There is an injective ring homomorphism $\varphi \colon R \to R[x]$ whose image is the set of constant polynomials.*

Proposition 1.3.33. *Let R and S be rings and let $\varphi \colon R \to S$ be a non-zero homomorphism. Suppose that R has a multiplicative identity 1_R. Then $\varphi(1_R)$ is a multiplicative identity in $\operatorname{Im} \varphi$.*

1.3.5 Quotient rings

Similarly, we need to enhance our definition of a normal subgroup appropriately in order to obtain something that behaves appropriately in giving us quotient rings (and not just quotients as groups). In more fancy language, we must work[5] out the necessary and sufficient conditions to define kernels in a suitable category.

Definition 1.3.34. *Let $(R, +, \times)$ be a ring. A subset $I \subseteq R$ is said to be an ideal if*

(I1) *I is an (additive) subgroup of $(R, +)$; and*

(I2) *(Closure under multiplication by an arbitrary element of R) for all $a \in I$ and $r \in R$, $a \times r \in I$ and $r \times a \in I$.*

If I is an ideal of R, we write $I \trianglelefteq R$; we write $I \triangleleft R$ if I is a proper ideal, that is, $I \trianglelefteq R$ and $I \subsetneq R$.

Proposition 1.3.35. *Let $(R, +, \times)$ be a ring and I an ideal of R.*
Let $L_I = \{r + I \mid r \in R\}$ denote the set of left cosets of I in R. Then

$$+_I \colon L_I \times L_I \to L_I, \quad (r + I) +_I (s + I) = (r + s) + I$$

and

$$\times_I \colon L_I \times L_I \to L_I, \quad (r + I) \times_I (s + I) = (r \times s) + I$$

define binary operations on L_I.

Proposition 1.3.36. *Let $(R, +, \times)$ be a ring and I an ideal.*

(i) *The operation $+_I$ defines a group structure on L_I, the set of left cosets of I in R (with respect to addition).*

(ii) *The operation $\times_I$ is associative and distributes over $+_I$.*

---

[5]This is blatantly ahistorical: the correct definition of a kernel in general categories is abstracted from the examples known already, i.e. groups, rings and other algebraic structures. But we are sowing seeds of a narrative that will play out later on.

Definition 1.3.37. Let $(R, +, \times)$ be a ring and let I be an ideal of R. The quotient ring of R by I, denoted R/I, is the ring $(L_I, +_I, \times_I)$ where L_I is the set $\{r + I \mid r \in R\}$ of left cosets of I in R and $+_I$ and $\times_I$ are the binary operations

$$(r + I) +_I (s + I) = (r + s) + I$$

and

$$(r + I) \times_I (s + I) = (r \times s) + I$$

The additive identity of the quotient ring R/I is $0_{R/I} = 0_R + I$.

Definition 1.3.38. Let $n \in \mathbb{Z}$. The ring of integers modulo n is defined to be the quotient ring $(\mathbb{Z}/n\mathbb{Z}, +_n, \times_n)$. We also often denote this ring by $(\mathbb{Z}_n, +_n, \times_n)$. We write $\widehat{m}$ for the coset $m + n\mathbb{Z}$, for short.

Proposition 1.3.39. *Let I be an ideal in a ring R.*

(i) *Suppose that R is commutative. Then the quotient ring R/I is commutative.*

(ii) *Suppose that R has a multiplicative identity 1_R. Then $1 + I$ is a multiplicative identity in the quotient ring R/I.*

1.3.6   The First Isomorphism Theorem for Rings

Now we have all the pieces to re-run the First Isomorphism Theorem story, but now for rings.

Proposition 1.3.40. *Let $\varphi \colon R \to S$ be a ring homomorphism. Then the kernel of φ, $\operatorname{Ker} \varphi$, is an ideal of R.*

Lemma 1.3.41. *Let R be a ring and I an ideal of R. The function $\pi \colon R \to R/I$ defined by $\pi(r) = r + I$ is a ring homomorphism.*

Definition 1.3.42. Let R be a ring and I an ideal of R. The ring homomorphism $\pi \colon R \to R/I$ defined by $\pi(r) = r + I$ is called the quotient homomorphism associated to R and I.

Proposition 1.3.43. *Let R be a ring and I an ideal of R. The quotient homomorphism $\pi \colon R \to R/I$ is surjective and its kernel is $\operatorname{Ker} \pi = I$.*

Corollary 1.3.44. *Let R be a ring and I an ideal of R. Then I is the kernel of a ring homomorphism with domain R, namely the quotient homomorphism $\pi \colon R \to R/I$.*

Theorem 1.3.45 (Universal property of the quotient ring). *Let R and S be rings. Let I be an ideal of R and let $\pi \colon R \to R/I$ be the associated quotient homomorphism.*

Then for every ring homomorphism $\varphi \colon R \to S$ such that $I \subseteq \operatorname{Ker} \varphi$, there exists a unique ring homomorphism $\bar{\varphi} \colon R/I \to S$ such that $\varphi = \bar{\varphi} \circ \pi$.

Furthermore, $\operatorname{Ker} \bar{\varphi} = (\operatorname{Ker} \varphi)/I$ and $\operatorname{Im} \bar{\varphi} = \operatorname{Im} \varphi$.

Theorem 1.3.46 (First Isomorphism Theorem for Rings). *Let R and S be rings and let $\varphi \colon R \to S$ be a ring homomorphism. Then*

$$R/\operatorname{Ker} \varphi \cong \operatorname{Im} \varphi.$$

1.3.7 Integral domains

Certain commutative rings are especially well-behaved.

Definition 1.3.47. An *integral domain* (or a *domain* for short) is a ring R such that the following two axioms hold.

(ID1) R is commutative and has a multiplicative identity.

(ID2) Whenever $a, b \in R$ are non-zero, the product ab is also non-zero.

Note that, by contraposition, axiom (ID2) is equivalent to the following statement.

(ID2') Whenever $a, b \in R$ satisfy $ab = 0$, either $a = 0$ or $b = 0$ (or both).

Example 1.3.48.

(a) The ring of integers, $\mathbb{Z}$, is an integral domain, and is regarded as the prototypical example, hence the term "integral". This is because the product of two non-zero integers is non-zero (and because $\mathbb{Z}$ is a commutative ring with a multiplicative identity).

(b) The ring $\mathbb{Z}_5$ is an integral domain. Certainly, $\mathbb{Z}_5$ is commutative and has a multiplicative identity, so that (ID1) holds.

To see that (ID2') is satisfied, suppose that $\widehat{m}, \widehat{n} \in \mathbb{Z}_5$ with $\widehat{m}\,\widehat{n} = \widehat{0}$. This means that $mn \equiv 0 \bmod 5$, so that 5 divides mn. Since 5 is prime, the only way mn can be a multiple of 5 is if either m or n is.

(This follows easily from Euclid's Lemma.)

Hence 5 divides m or 5 divides n, and thus $\widehat{m} = \widehat{0}$ or $\widehat{n} = \widehat{0}$, as required.

More generally, $\mathbb{Z}_p$ is an integral domain for each prime $p \in \mathbb{N}$; the proof is the same as that above.

(c) Other familiar examples of integral domains are $\mathbb{Q}, \mathbb{R}$ and $\mathbb{C}$. The consideration of these cases is similar to that of $\mathbb{Z}$ above.

Example 1.3.49. The ring $\mathbb{Z}_6$ is *not* an integral domain because $\widehat{2} \cdot \widehat{3} = \widehat{0}$ in $\mathbb{Z}_6$ and both $\widehat{2}$ and $\widehat{3}$ are non-zero, so that (ID2) does not hold.

More generally, the ring $\mathbb{Z}_n$ is not an integral domain whenever $n \in \mathbb{N}$ is composite.

Definition 1.3.50. A non-zero element a of a commutative ring R is called a *zero-divisor* if there exists a non-zero element $b \in R$ such that $ab = 0$.

Thus a commutative ring with a multiplicative identity is an integral domain if and only if it has no zero-divisors.

Example 1.3.51. We find the zero-divisors in the ring $\mathbb{Z}_6$ (there are three of them).

We consider each of the six elements of $\mathbb{Z}_6$ separately.

- The zero element $\widehat{0}$ is not a zero-divisor by definition.

- The element $\widehat{1}$ satisfies $\widehat{1}\,\widehat{m} = \widehat{m}$ for each $m \in \mathbb{Z}$ so we see that $\widehat{1}\,\widehat{m} = \widehat{0}$ only for $\widehat{m} = \widehat{0}$, so that $\widehat{1}$ is not a zero-divisor.

  (In general, a multiplicative identity is never a zero-divisor.)

- The element $\widehat{2}$ is a zero-divisor because $\widehat{2} \neq \widehat{0}$ and $\widehat{3} \neq \widehat{0}$, and $\widehat{2} \cdot \widehat{3} = \widehat{0}$.

- The element $\widehat{3}$ is also a zero-divisor by the argument just given.

- The element $\widehat{4}$ is a zero-divisor because $\widehat{4} \neq \widehat{0}$ and $\widehat{3} \neq \widehat{0}$, and $\widehat{4} \cdot \widehat{3} = \widehat{0}$.

- The element $\widehat{5}$ is not a zero-divisor. Indeed, suppose that $m \in \mathbb{Z}$ satisfies $\widehat{5}\,\widehat{m} = \widehat{0}$. Then, as $\widehat{5}\,\widehat{m} = \widehat{5m}$, this means that 6 divides $5m$, and consequently 6 divides m because 5 and 6 are coprime. Thus $\widehat{m} = \widehat{0}$, as required.

We could get some help from SageMath to do this, as follows.

```
1  Z6 = Integers(6)
2  list(Z6)
3  Z6(2)*Z6(3)
4  l = []
5  for i in range(6):
6          l.append(Z6(5)*Z6(i))
7  l
```

Here we have used a for loop for the first time. Note that since SageMath follows Python conventions, the indentation (achieved by pressing Tab once) on line 6 is essential.

The last four lines do the following: define an empty list l, ask SageMath to range a variable i between 0 and 5 (range(n) goes from 0 to n−1) and for each i put on the end of the list so far ("append") the value of Z6(5)*Z6(i), finally printing the result.

Why does the output of this code confirm that $\widehat{5}$ is not a zero divisor?

A useful property of integral domains is that they allow multiplicative cancellation of non-zero elements. Consider for example the equation $2m = 2n$, where m and n are integers. Although 2 has no multiplicative inverse in $\mathbb{Z}$, it is still permissible to cancel the 2 from both sides to obtain $m = n$. This property can be generalized to an arbitrary integral domain as follows.

Proposition 1.3.52. *Let R be a commutative ring with a multiplicative identity. Then R is an integral domain if and only if multiplicative cancellation is always possible in the following sense.*

(MC) *Whenever $a, b, c \in R$ satisfy $ab = ac$ and $a \neq 0$, we have $b = c$.*

Definition 1.3.53. Let R be a commutative ring. A *prime ideal* in R is a subset I of R such that the following two axioms hold.

(PI1) I is a proper ideal in R (so $I \neq R$).

(PI2) Whenever $a, b \in R$ satisfy $ab \in I$, either $a \in I$ or $b \in I$ (or both).

Proposition 1.3.54. *Let R be a commutative ring with a multiplicative identity, and let I be an ideal in R. The quotient ring R/I is an integral domain if and only if I is a prime ideal.*

1.3.8   Fields

Although we have come to them after groups and rings, it is often fields that we meet first in our mathematical journey. We might not know the name, or see the justification for the axioms, but we learn early on that the real numbers have lots of nice properties, like being able to carry out division.

Definition 1.3.55. A *field* is a ring F such that the following two axioms hold.

(F1) F is commutative and has a multiplicative identity 1.

(F2) Each non-zero element of F has a multiplicative inverse, that is, for each element $a \in F \setminus \{0\}$, there is an element $b \in F$ such that $ab = 1$.

Using the language of group theory, we can rephrase the field axioms as follows.

Proposition 1.3.56. *Let F be a non-empty set equipped with two binary operations $+$ and $\times$. Then F is a field if and only if the following three conditions hold.*

(i) *$(F, +)$ is an Abelian group.*

(ii) *$(F \setminus \{0\}, \times)$ is an Abelian group.*

(iii) *The distributive laws hold.*

Proposition 1.3.57. *Let F be a field. Then F is an integral domain.*

Proposition 1.3.58. *Let R be an integral domain with a finite number of elements. Then R is a field.*

Although we will mostly have infinite fields in mind in the remainder of the book (indeed, most examples will be over $\mathbb{C}$), the following serves as a useful orientation and connection of the various definitions we have seen.

Corollary 1.3.59. *For each $n \in \{2, 3, 4, \ldots\}$, the following four assertions are equivalent.*

(i) $\mathbb{Z}_n$ *is a field.*

(ii) $\mathbb{Z}_n$ *is an integral domain.*

(iii) *n is prime.*

(iv) *$n\mathbb{Z}$ is a prime ideal in $\mathbb{Z}$.*

In SageMath, the fields $\mathbb{Z}_p$ are called GF(p) and work similarly[b] to Integers; the difference is that SageMath "remembers more is true" about GF(p).

```
1   Z5 = GF(5)
2   Z5(2)*Z5(4)
```

---

[b]Warning: GF(n) is a valid construction if n is a prime power, giving the finite field of that order, and will return an error if n is not a prime power. So, use Integers if you want integers modulo n, not GF, or the results may not be what you expect!

Note that by adding a requirement to be closed under taking multiplicative inverses, we obtain the notion of a subfield, exactly analogously to subgroups and subrings.

We say that that a ring R is simple if the two trivial ideals $\{0\}$ and R are the *only* ideals in R. We have that, among all unital commutative rings, fields are characterized as those which are simple.

Proposition 1.3.60. *Let R be a ring. Then R is a field if and only if R is simple and commutative and has a multiplicative identity.*

Corollary 1.3.61. *Let R be a field and $\varphi\colon R \to S$ a ring homomorphism. Then either $\varphi = 0$ or φ is injective.*

As an application of Proposition 1.3.60 we can characterize the ideals I in a commutative unital ring R such that the quotient ring R/I is a field; this is the analogue for fields of Proposition 1.3.54 for integral domains.

Definition 1.3.62. *Let R be a ring. A maximal ideal in R is a subset I of R such that the following two axioms hold.*

(MI1) *I is a proper ideal in R.*

(MI2) *Suppose that J is an ideal in R such that $I \subseteq J$. Then either $J = I$ or $J = R$.*

In other words, a maximal ideal is a proper ideal which is not properly contained in any other proper ideal. In applications, the following more compact reformulation of axiom (MI2) is often convenient.

(MI2′) Suppose that J is an ideal in R such that $I \subsetneq J$ (meaning that $I \subseteq J$ and $I \neq J$). Then $J = R$.

Proposition 1.3.63. *Let R be a commutative ring with a multiplicative identity, and let I be an ideal in R. The quotient ring R/I is a field if and only if I is a maximal ideal in R.*

Corollary 1.3.64. *Let R be a commutative ring with a multiplicative identity. Then every maximal ideal in R is a prime ideal.*

1.3.9 Principal ideal domains

We have seen that $n\mathbb{Z}$ is an ideal in $\mathbb{Z}$ for each $n \in \mathbb{Z}$. This raises the question: are there any other ideals in $\mathbb{Z}$? The answer is "no", as we shall see shortly.

Proposition 1.3.65. *Let a be an element of a commutative ring R.*

(i) *The set $aR = \{ab : b \in R\}$ is an ideal in R.*

(ii) *Suppose that R has a multiplicative identity 1. Then $a \in aR$, and aR is the smallest ideal in R containing a.*

Definition 1.3.66. An ideal I in a commutative ring R is called principal if $I = aR$ for some $a \in R$.

Definition 1.3.67. A *principal ideal domain* (often abbreviated *PID*) is a ring R such that the following two axioms hold.

(PID1) R is an integral domain.

(PID2) Each ideal in R is principal.

In other words, an integral domain R is a principal ideal domain if and only if, for each ideal I in R, there exists an element $a \in R$ such that $I = aR$.

We begin with an easy (and admittedly rather uninteresting) example of a principal ideal domain.

Proposition 1.3.68. *Let F be a field. Then F is a principal ideal domain.*

A much more important example of a principal ideal domain is as follows.

Theorem 1.3.69. *The ring of integers, $\mathbb{Z}$, is a principal ideal domain.*

We have the following nice description of the ideal structure of $\mathbb{Z}$.

Corollary 1.3.70.

(i) *Each ideal in $\mathbb{Z}$ has the form $n\mathbb{Z}$ for some $n \in \mathbb{Z}$.*

(ii) *Let $m\mathbb{Z}$ and $n\mathbb{Z}$ be two ideals in $\mathbb{Z}$ (where $m, n \in \mathbb{Z}$). Then $m\mathbb{Z} \subseteq n\mathbb{Z}$ if and only if n divides m; in particular, $m\mathbb{Z} = n\mathbb{Z}$ if and only if $m = \pm n$.*

Proposition 1.3.71. *Let $m, n \in \mathbb{N}$. Then:*

(i) *The smallest ideal in $\mathbb{Z}$ containing both $m\mathbb{Z}$ and $n\mathbb{Z}$ is $\mathrm{hcf}(m, n)\mathbb{Z}$.*

(ii) *The largest ideal in $\mathbb{Z}$ contained in both $m\mathbb{Z}$ and $n\mathbb{Z}$ is $\mathrm{lcm}(m, n)\mathbb{Z}$ (where $\mathrm{lcm}(m, n)$ denotes the lowest common multiple of m and n).*

The following is an illustration of the lattice of ideals of $\mathbb{Z}$:

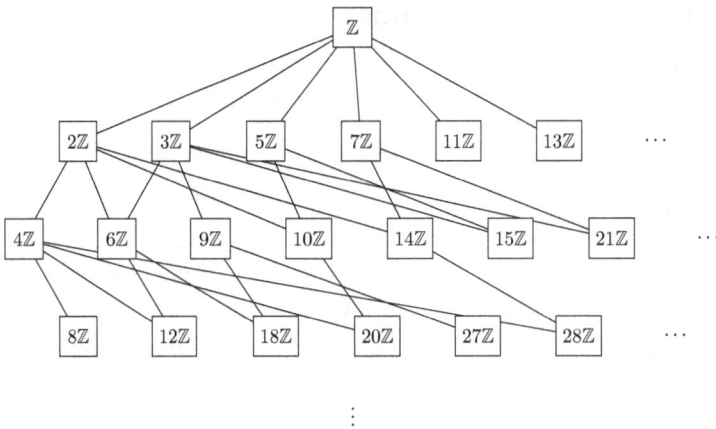

Using the definition of the degree of a polynomial and the division algorithm, one may prove the following.

Proposition 1.3.72. *Let R be a ring. Then $R[x]$ is an integral domain if and only if R is an integral domain.*

Theorem 1.3.73. *The polynomial ring $F[x]$ over a field F is a principal ideal domain.*

1.4   Linear algebra

The word "linear" has several meanings, but in common usage it derives its main one from the notion of a line: we say a geographical feature is linear if it is (perhaps only roughly) in the shape of a (straight) line. Mathematically, we use linear in this way too, notably in geometry, but in algebra there is an arguably more important usage that is one level more sophisticated and is another instance of wanting functions to preserve structure.

We say that a function is linear if it sends lines (through the origin) to lines. More concretely, f will satisfy $f(\alpha v + \beta w) = \alpha f(v) + \beta f(w)$, where α, β will be "scalars" and v, w points (or "vectors") in some suitable setting. Functions that are linear are considerably easier to work with than non-linear functions, as is seen in applications across mathematics, physics and engineering. Indeed, many successful techniques for solving problems are essentially "how can we make things linear?" and this is why linear algebra is often taught right at the beginning of a course in advanced mathematics (and some aspects even earlier).

The "suitable" setting referred to above is, at least from the point of view of abstract algebra, that of vector spaces. We very briefly mentioned vector spaces in our introductory remarks in Section 1.1 and now let us say where they fit in terms of the formal definitions.

Every vector space V over a field $\mathbb{K}$ is an Abelian group under addition, $(V, +)$. Vector spaces are not rings (or fields): we do not multiply vectors by vectors.

However, vector spaces have more structure than just addition, namely they have scalar multiplication. Scalar multiplication is a slightly different type of operation: it is not a binary operation on V. It takes two inputs, a scalar $\lambda \in \mathbb{K}$ and a vector $v \in V$, and produces a new vector $\lambda v \in V$. We can formalize this as a function $\cdot : \mathbb{K} \times V \to V, \lambda \cdot v = \lambda v$, and say that a vector space is an algebraic structure $(V, +, \cdot)$ where $+$ and $\cdot$ satisfy some compatibility conditions, namely those that give linearity.

With full formality, the definition is as follows.

Definition 1.4.1. A vector space over a field $\mathbb{K}$ is an algebraic structure $(V, +, \cdot)$ such that $(V, +)$ is an Abelian group with identity 0_V and the function $\cdot : \mathbb{K} \times V \to V$ satisfies

$$\lambda \cdot (v + w) = \lambda \cdot v + \lambda \cdot w,$$
$$(\lambda + \mu) \cdot v = \lambda \cdot v + \mu \cdot v,$$
$$\lambda \cdot (\mu \cdot v) = (\lambda \mu) \cdot v$$

and

$$1_{\mathbb{K}} \cdot v = v$$

for all $v, w \in V$ and $\lambda, \mu \in \mathbb{K}$.

This definition is a little more complicated than our previous examples, since for a vector space we should regard $\mathbb{K}$ (itself an algebraic structure with operations, distinguished elements and properties) as part of the data we pick. As a result, the theory of vector spaces is quite rich and indeed it gets its own special name, linear algebra.

Since much of representation theory is both built on and inspired by linear algebra, we take a little time here to discuss the most relevant parts.

Again, we note that there is much more we could include, e.g. the theory of bilinear forms (an example of a topic that has applications in many areas), but which we omit—with a mild sense of regret—in order to stay on our main track.

To define the usual algebraic constructions for vector spaces, we again simply "enhance" the group-theoretic concepts to vector spaces by asking or checking at each stage that the idea in question can be made compatible with the extra scalar multiplication, as we did with rings.

Specifically, a *subspace* W of a vector space V is a subgroup of V closed under scalar multiplication.

A homomorphism of vector spaces or, to give it its more common name, a *linear transformation* (or map) $T: V \to W$ is a homomorphism of Abelian groups such that $\varphi(\lambda v) = \lambda \varphi(w)$. Let us denote by $\text{Hom}_{\mathbb{K}}(V, W)$ the collection of all linear transformations from V to W.

As a nod to how we will think about structures like vector spaces in the rest of this book, notice that the scalar multiplication formula $\cdot$ could be re-interpreted as a function $\rho.: \mathbb{K} \to \text{Hom}_{\mathbb{K}}(V, V)$, given by $\rho.(\lambda)(v) = \lambda \cdot v$. Then the axioms for $\cdot$ precisely say that $\rho.$ is a unital ring homomorphism.

We note here that most of the time (but not all) the vector spaces we will encounter will be *finite-dimensional*. However, we will take care to say so specifically, as the theory of infinite-dimensional vector spaces comes with some subtleties.

Two important quantities associated to a linear endomorphism $T: V \to V$ of a finite-dimensional vector space are the determinant and the trace. To define both,[6] we choose a *basis* $\mathcal{B}$ for V. Then we may write down the *matrix* $[T]_{\mathcal{B}}^{\mathcal{B}} = (t_{ij})$ whose entries are determined by $Tb_i = \sum_j t_{ij} b_j$ where $b_i \in \mathcal{B}$. The *trace* of T, $\text{tr}(T)$ or $\text{tr}([T]_{\mathcal{B}}^{\mathcal{B}})$, is the sum of the diagonal entries of $[T]_{\mathcal{B}}^{\mathcal{B}}$, i.e. $\sum_i t_{ii}$; one should show that this does not depend on the choice of basis.

The *determinant* is a little more complicated:

$$\det(T) \stackrel{\text{def}}{=} \sum_{\sigma \in S_n} \text{sign}(\sigma) t_{1\sigma(1)} t_{2\sigma(2)} \cdots t_{n\sigma(n)}$$

where $[T]_{\mathcal{B}}^{\mathcal{B}} = (t_{ij})$; recall (1.2.2) that the sign of a permutation σ is $(-1)^r$ if σ can be written as a product of r transpositions.

Eigenvectors and eigenvalues will be important for us, too. An *eigenvector* for a linear map $T: V \to V$ is a vector $v \in V$ such that there exists $\lambda \in \mathbb{K}$ such that $Tv = \lambda v$. The scalar λ is called the *eigenvalue* associated to the eigenvector v.

To find eigenvalues, we may use the following result.

---

[6]To introduce all the required definitions and theorems would take more space than we wish to here: in order to follow this book, we expect that you have seen this type of abstract linear algebra before. If not, we recommend you find a good textbook to work through first; there are many available, to suit a variety of backgrounds.

Theorem 1.4.2. *Let* $T: V \to V$ *and set* $M = [T]_{\mathcal{B}}^{\mathcal{B}}$ *for some choice of basis $\mathcal{B}$ for* V. *Then for any* $\lambda \in \mathbb{K}$ *there exists an eigenvector v for T with eigenvalue λ if and only if λ is a root of the characteristic polynomial of T,* $\det(M - tI)$.

Then to find eigenvectors, we solve the simultaneous linear equations $Mv = \lambda v$. (Fast computational methods to compute eigenvectors and eigenvalues exist but it is also important to know how to do this by hand.)

The subset of eigenvectors for a fixed eigenvalue λ forms a subspace of V, which we denote V_λ and call the λ-*eigenspace*.

If a vector space V has a basis consisting of eigenvectors for a linear transformation $T: V \to V$, then we say T is *diagonalisable*, for then with respect to this basis, T is represented by a diagonal matrix with the eigenvalues on the diagonal. It follows that if T is diagonalisable, its trace is equal to the sum of its eigenvalues and its determinant is equal to the product of its eigenvalues.

We now show you how to define matrices in SageMath and then to compute their eigenvalues and eigenvectors. In the exercises (1.E) at the end of this chapter (and indeed, the other chapters too) there are opportunities to practice these computations.

Putting the code

```
1  R = MatrixSpace(ZZ,3,3)
2  M = R.matrix([[1,2,3],[4,5,6],[7,8,9]])
3  M
```

into SageMathCell and pressing "Evaluate" prints

$$\begin{bmatrix} 1 & 2 & 3 \\ 4 & 5 & 6 \\ 7 & 8 & 9 \end{bmatrix}$$

So now we know how to create the matrix ring $M_3(\mathbb{Z})$ and how to define M to be a matrix in this ring.

We can add matrices:

```
4  M+M
```

and multiply matrices:

```
5  M*M
```

For what follows, and for going beyond what follows, you might find the SageMath tutorial on linear algebra (https://doc.sagemath.org/html/en/tutorial/tour_linalg.html) or the documentation for Matrices (https://doc.sagemath.org/pdf/en/reference/matrices/matrices.pdf) helpful, depending on what you want to do. The technical documentation can be daunting at first, but often the examples help you see how to achieve what you want.

It should be said again at this point that computers struggle with infinite things. This ought not to come as a surprise but it does matter. It means, in particular, that in the examples that follow immediately and later we will pick examples that the system can handle. Also, the outputs will look approximate in many cases, although we might be able to infer expressions for them by knowing that they are (for example) roots of certain polynomials whose coefficients are manageable for the system (which for our purposes essentially means "rational").

Computing eigenvalues is as you might expect:

```
1  R = MatrixSpace(ZZ,3,3);
2  M = R.matrix([[1,2,3],[4,5,6],[7,8,9]]);
3  M.eigenvalues()
```

yields as output [0, -1.116843969807043?, 16.11684396980705?]. By asking for the characteristic polynomial of M, we can work out what these numbers might actually be:

```
4  M.characteristic_polynomial().factor()
```

We see that these are the roots of $x(x^2 - 15x - 18)$, i.e. 0 and $\frac{1}{2}(15 \pm 3\sqrt{33})$.

Moreover, we can compute the eigenvectors:

```
5  M.eigenvectors_right()
```

gives

```
[(0, [(1, -2, 1)], 1),

(-1.116843969807043?, [(1, 0.110394503377411963?,

-0.7792109924517608?)], 1),

(16.11684396980705?, [(1, 2.264605496225881?,

3.529210992451761?)], 1)]
```

i.e. a list of lists whose entries are the eigenvalue, an eigenvector and its algebraic multiplicity.

1.4.1 Advanced linear algebra

We need two linear algebra constructions that go beyond a typical first course. These are quotients for vector spaces and tensor products.

Quotient vector spaces

Since vector spaces are Abelian groups and every subgroup of an Abelian group is a normal subgroup, we may take the quotient group and show that this inherits a vector space structure.

Let W be a subspace of V. Then the *quotient vector space* V/W is the Abelian group $V/W = \{v + W \mid v \in V\}$ with $(v + W) + (v' + W) = (v + v') + W$ and $\lambda \cdot (v + W) = \lambda v + W$. We leave it as an exercise to check the necessary properties; they all "descend" from those for V. (The well-definedness of the construction has already been handled by the theory of quotient groups.)

Note that if V is a finite-dimensional vector space and W a subspace of V, then $\dim V/W = \dim V - \dim W$. Indeed this follows from proving the stronger result that $V \cong W \oplus V/W$.

The point here is that due to the "freeness" of vector spaces (essentially, this boils down to the existence of bases), in the category of vector spaces every short exact sequence of vector spaces splits. The correct, sophisticated way to say this is that $\mathbb{K}$-Mod is a semisimple category.[7]

Tensor products of vector spaces

Our definition of the tensor product of two vector spaces will be given by formalizing the following. Given V, W $\mathbb{K}$-vector spaces, their Cartesian product $V \times W$ is a vector space and it is natural to consider bilinear maps $h\colon V \times W \to Z$. However, bilinear maps can be awkward to work with. We would like to be able to use linear algebra results, but these are framed in terms of linear maps. A resolution is to construct a new $\mathbb{K}$-vector space, the tensor product $V \otimes W$, such that h can always be replaced by a *linear* map from $V \otimes W$ to Z.

Definition 1.4.3. Let V and W be $\mathbb{K}$-vector spaces. The pair $(X, \otimes)$ is said to be the *tensor product* of V and W if

(a) X is a $\mathbb{K}$-vector space;

(b) $\otimes\colon V \times W \to X$ is a bilinear map; and

(c) (universal property; 2.4) for every vector space Z and bilinear map $h\colon V \times W \to Z$, there exists a unique linear map $\tilde{h}\colon X \to Z$ such that $h = \tilde{h} \circ \otimes$.

$$
\begin{array}{ccc}
V \times W & \xrightarrow{\ \otimes\ } & X \\
& h \searrow & \big\downarrow \tilde{h} \\
& & Z
\end{array}
$$

That is, every bilinear map $h\colon V \times W \to Z$ *factors through* $\otimes$. A significant advantage of this definition is that the following is immediate.

---

[7]Several of the terms in this paragraph are yet to be defined; see Chapters 2 and 4 in particular.

Lemma 1.4.4. *The vector space X is uniquely determined up to unique isomorphism.*

Proof. If $(X, \otimes)$ and $(X', \otimes')$ are pairs satisfying the conditions of the definition, putting $Z = X'$ we have $\otimes' = \tilde{\otimes}' \circ \otimes$ and conversely via the universal property for $\otimes'$, $\otimes = \tilde{\otimes} \circ \otimes'$ for unique maps $\tilde{\otimes}$ and $\tilde{\otimes}'$.

We therefore have a commuting diagram (2.2.2):

$$
\begin{array}{ccc}
V \times W & \xrightarrow{\ \otimes\ } & X \\
\ \downarrow{\scriptstyle \mathrm{id}} & \searrow{\scriptstyle \otimes'} & \ \ \vdots\ {\scriptstyle \tilde{\otimes}'} \\
V \times W & \xrightarrow{\ \otimes'\ } & X' \\
\ \downarrow{\scriptstyle \mathrm{id}} & \searrow{\scriptstyle \otimes} & \ \ \vdots\ {\scriptstyle \tilde{\otimes}} \\
V \times W & \xrightarrow{\ \otimes\ } & X
\end{array}
$$

Examining the outside rectangle, we see that we in fact have a triangle

$$
\begin{array}{ccc}
V \times W & \xrightarrow{\ \otimes\ } & X \\
 & \searrow{\scriptstyle \otimes} & \ \ \vdots\ {\scriptstyle \tilde{\otimes} \circ \tilde{\otimes}'} \\
 & & X
\end{array}
$$

in which $\otimes = (\tilde{\otimes} \circ \tilde{\otimes}') \circ \otimes$. But id_X is also a map making the triangle commute, so by the universal property, $\tilde{\otimes} \circ \tilde{\otimes}' = \mathrm{id}_X$. By the corresponding argument with X and X', and $\tilde{\otimes}$ and $\tilde{\otimes}'$, interchanged, we conclude that $\tilde{\otimes}' \circ \tilde{\otimes} = \mathrm{id}_{X'}$ also and hence that $\tilde{\otimes}'$ and $\tilde{\otimes}$ are inverse isomorphisms between X and X'. □

We can go further and use the universal property to prove the following.

Lemma 1.4.5. *Let V and W be $\mathbb{K}$-vector spaces and let $(X, \otimes)$ (respectively $(Y, \otimes')$) be the tensor product of V and W (respectively W and V). Then $X \cong Y$ as $\mathbb{K}$-vector spaces.*

Proof. Let $\tau \colon V \times W \to W \times V$ be the function $\tau(v, w) = (w, v)$. Note that τ is bilinear: for example,

$$
\begin{aligned}
\tau(v_1 + v_2, w) &= (w, v_1 + v_2) \\
&= (w, v_1) + (w, v_2) \\
&= \tau(v_1, w) + \tau(v_2, w).
\end{aligned}
$$

Furthermore, τ is an isomorphism, with $\tau^{-1}(w, v) = (v, w)$.

The remainder of the argument closely parallels that of the proof of the previous lemma. Let $h \colon V \times W \to Y$ be the bilinear map $\otimes' \circ \tau$. By the universal property, there exists a unique linear map $\tilde{h} \colon X \to Y$ such that $h = \tilde{h} \circ \otimes$. Similarly, letting $k \colon W \times V \to X$, $k = \otimes \circ \tau^{-1}$ we obtain $k = \tilde{k} \circ \otimes'$.

Consider $\tilde{k} \circ \tilde{h} \colon X \to X$. We have the following diagram, which commutes by the above:

$$
\begin{array}{ccc}
V \times W & \xrightarrow{\;\otimes\;} & X \\
\Big\downarrow{\scriptstyle \tau} & \overset{h}{\searrow} & \Big\downarrow{\scriptstyle \tilde{h}} \\
W \times V & \xrightarrow{\;\otimes'\;} & Y \\
\Big\downarrow{\scriptstyle \tau^{-1}} & \overset{k}{\searrow} & \Big\downarrow{\scriptstyle \tilde{k}} \\
V \times W & \xrightarrow{\;\otimes\;} & X
\end{array}
$$

Examining the outside rectangle, we see that we in fact have a triangle

$$
\begin{array}{ccc}
V \times W & \xrightarrow{\;\otimes\;} & X \\
& \overset{\otimes}{\searrow} & \Big\downarrow{\scriptstyle \tilde{k}\circ\tilde{h}} \\
& & X
\end{array}
$$

in which $\otimes = \tilde{k} \circ \tilde{h} \circ \otimes$. But id_X is also a map making the triangle commute, so by the universal property, $\tilde{k} \circ \tilde{h} = \mathrm{id}_X$. By the corresponding argument with X and Y, and $\tilde{h}$ and $\tilde{k}$, interchanged, we conclude that $\tilde{h} \circ \tilde{k} = \mathrm{id}_Y$ also and hence that $\tilde{h}$ and $\tilde{k}$ are inverse isomorphisms between X and Y. □

A significant *disadvantage* of our definition of a tensor product is that it is not immediate that such a pair $(X, \otimes)$ exists. This is frequently the case with universal property definitions: one also needs to construct a *model*. (One model suffices, since all models will be isomorphic, up to unique isomorphism.)

The standard model of the tensor product is given by taking the vector space spanned by all symbols $v \otimes w$ with $v \in V$ and $w \in W$, and then imposing on this the relations which give $\otimes$ the bilinearity properties we want. More formally let T denote the vector space spanned by the set of symbols $\{v \otimes w \mid v \in V, w \in W\}$. Let I be the subspace of T spanned by the following elements:

$$
\begin{aligned}
& (v_1 + v_2) \otimes w - v_1 \otimes w - v_2 \otimes w \\
& v \otimes (w_1 + w_2) - v \otimes w_1 - v \otimes w_2 \\
& (\lambda v) \otimes w - \lambda(v \otimes w) \\
& v \otimes (\mu w) - \mu(v \otimes w)
\end{aligned}
$$

where $v, v_1, v_2 \in V$, $w, w_1, w_2 \in W$ and $\lambda, \mu \in \mathbb{K}$. Then define $V \otimes W = T/I$, the quotient vector space. We abuse/overload notation by writing $v \otimes w$ for $v \otimes w + I$.

Then the above lemma shows that there exists an isomorphism

$$
\tau \colon V \otimes W \to W \otimes V, \quad \tau(v \otimes w) = w \otimes v
$$

(this being the isomorphism denoted by $\tilde{h}$ in the proof; with that result in hand, it is natural to abuse notation and also call this map τ).

It is not hard to show that if $\mathcal{B}_V$ is a basis for V and $\mathcal{B}_W$ a basis for W, a basis for $V \otimes W$ is given by $\{b \otimes c \mid b \in \mathcal{B}_V, c \in \mathcal{B}_W\}$.

Examples 1.4.6. Let V be a vector space over $\mathbb{K}$.

(a) If $\dim V = 1$ and $\mathcal{B} = \{v\}$ is a basis, then $V \otimes V = \text{span}_{\mathbb{K}}\{v \otimes v\}$ is also 1-dimensional.

(b) If $\dim V = 2$ and $\mathcal{B} = \{v, w\}$ is a basis, then

$$V \otimes V = \text{span}_{\mathbb{K}}\{v \otimes v, v \otimes w, w \otimes v, w \otimes w\}$$

is 4-dimensional.

(c) If $\dim V = 2$ and $\mathcal{B}_V = \{v, w\}$ is a basis and X is a 1-dimensional vector space with basis $\mathcal{B}_X = \{x\}$, then

$$V \otimes X = \text{span}_{\mathbb{K}}\{v \otimes x, w \otimes x\}$$

is 2-dimensional.

Indeed, if V and W are finite-dimensional, $\dim V \otimes W = \dim V \cdot \dim W$. (Recall by way of comparison that $\dim V \oplus W = \dim V + \dim W$.)

There is much more one could say about the tensor product but we will end by remarking that we have deliberately given a definition that extends, essentially without modification, from $\mathbb{K}$-modules (i.e. vector spaces) to R-modules for R a commutative unital ring (see Section 4 for the definition of an R-module).

We will examine tensor products and related ideas in more detail in Section 2.7 and Section 6.3.

1.5 Algebras

Our aim in this section is to introduce the notion of an *algebra*. We will simplify slightly and consider only algebras over fields at this point. Later, in Chapter 4, we will extend this to algebras over commutative unital rings.

The two main complementary ways to think about what an algebra is are "a vector space with an associative multiplication" and "a ring with an extra scalar multiplication by elements of a field".

When we first study linear algebra, it is often emphasized that *you can't multiply vectors*, in order to make the point that vector spaces are not rings or fields. This can lead to confusion: $\mathbb{R}$ is a (1-dimensional) vector space and we can multiply its vectors! But generally, the warning is the right one—away from the trivial case of dimension one, vector spaces are not (usually) rings or fields (in a natural way).[8]

Now that we have understood (i) that there are different types of algebraic structure, (ii) that some examples fall into more than one class and (iii) when they do, we should be clear about which we are thinking about, we can allow ourselves to ask and indeed answer the question, "well, what happens if we are allowed to multiply vectors?"

Definition 1.5.1. Let $\mathbb{K}$ be a field. We say that $A = (A, +, \cdot, \times)$ is an *associative unital $\mathbb{K}$-algebra* if A is a $\mathbb{K}$-vector space via $+$ and $\cdot$, A is also a unital ring with respect to $+$ and $\times$ and the two structures are compatible, i.e.

$$\lambda \cdot (a \times b) = (\lambda \cdot a) \times b = a \times (\lambda \cdot b)$$

for all $\lambda \in \mathbb{K}$, $a, b \in A$.

Since we will always assume our algebras are associative and unital, we will just say "$\mathbb{K}$-algebra" for short from now on.

Examples 1.5.2.

(a) $\mathbb{C}$ is an $\mathbb{R}$-algebra, since complex multiplication is $\mathbb{R}$-linear. Similarly, $\mathbb{R}$ is a $\mathbb{Q}$-algebra. The 'base' of this family of algebras (over fields) is $\mathbb{Q}$: all of $\mathbb{Q}$, $\mathbb{R}$ and $\mathbb{C}$ are $\mathbb{Q}$-algebras.[9]

(b) Every field $\mathbb{K}$ is a $\mathbb{K}$-algebra, taking $\cdot = \times$ (the multiplication in the field $\mathbb{K}$). That is, in 1-dimensional vector spaces, you can multiply the vectors (which are just scalars!).

(c) Polynomials naturally have an algebra structure: if $\mathbb{K}$ is a field, $\mathbb{K}[x]$ is a $\mathbb{K}$-algebra. The base field acts as the scalars for the polynomials: $\lambda \triangleright p(x)$ is just multiplication by λ, since we start from $\lambda \triangleright x^n = \lambda x^n$ and extend $\mathbb{K}$-linearly. So $\mathbb{R}[x]$ is an $\mathbb{R}$-algebra, for example.

(d) Matrices also form an algebra: $M_n(\mathbb{K})$ is a $\mathbb{K}$-algebra, where $\lambda \triangleright M$ is the 'usual' multiplication of a matrix by a scalar, i.e. multiply each entry by λ (or equivalently, $\lambda \triangleright M = \lambda I \cdot M$). So $M_n(\mathbb{C})$ is a $\mathbb{C}$-algebra, for example.

We also have the natural notion of algebra homomorphisms, i.e. maps between algebras preserving the algebra structure.

---

[8]It is hard for this author to let this discussion pass without saying the words "vector cross product" or "Lie algebra", but I promised not to talk about non-associative algebras. This footnote will have to suffice as a prompt to those who are interested to go and find out more elsewhere.

[9]The more general definition of an algebra over a commutative unital ring will allow us to reinstate $\mathbb{Z}$ in its natural place at the bottom of this hierarchy.

Definition 1.5.3. Let $(A, +_A, \cdot_A, \times_A)$ and $(B, +_B, \cdot_B, \times_B)$ be $\mathbb{K}$-algebras. A homomorphism of $\mathbb{K}$-algebras (or *algebra homomorphism*) is a function $f\colon A \to B$ such that

(a) (*vector space structure preserved*) f is a $\mathbb{K}$-linear transformation,

(b) (*multiplication preserved*) $f(a_1 \times_A a_2) = f(a_1) \times_B f(a_2)$ for all $a_1, a_2 \in A$, and

(c) (*multiplicative identity preserved*) $f(1_A) = 1_B$.

Notice that preservation of addition and scalar multiplication is bundled up in the first condition and that the second two along with the preservation of addition say that f is a (unital) homomorphism of unital rings.

Definition 1.5.4. Let $(A, +_A, \cdot_A, \times_A)$ and $(B, +_B, \cdot_B, \times_B)$ be $\mathbb{K}$-algebras. A homomorphism of $\mathbb{K}$-algebras $f\colon A \to B$ is said to be an *algebra isomorphism* if there exists an algebra homomorphism $g\colon B \to A$ such that $g \circ f = \mathrm{id}_A$ and $f \circ g = \mathrm{id}_B$.

As before, an algebra homomorphism is an isomorphism if and only if it is bijective.[10]

By now, it should hopefully be broadly clear how we should define *subalgebras*: these are subsets of a $\mathbb{K}$-algebra that themselves admit an algebra structure. There is a small subtlety, though, in that we need to insist that a subalgebra is an algebra over the *same* field; we are not allowed to mix and match. In practice, there are ways to handle this, however, via a construction called "extension of scalars".

Then one can easily check that a subalgebra needs to be both a vector subspace and a subring, and that these conditions suffice.

Since an algebra A is in particular a (unital) ring, it has *ideals I* and we can form the *quotient algebra A/I*. The compatibility of the ring and vector space structure means that this gives the same underlying set as taking the vector space quotient of A by I (which is in particular a subspace of A). Both the ring and vector space structures descend to the quotient, with compatibility preserved, so that A/I is again an algebra.

Problem 7. Let $A = \mathbb{C}[x]/I$ be the algebra given by the quotient of the polynomial algebra in one variable over $\mathbb{C}$, $\mathbb{C}[x]$, by the ideal $I = \langle x^5 - 1 \rangle$ generated by $x^5 - 1$. Using the fact that $\mathcal{B}' = \{1, x, x^2, x^3, \dots\}$ is a basis for $\mathbb{C}[x]$ and that hence the set $\{1 + I, x + I, x^2 + I, x^3 + I, \dots\}$ spans A, find a basis for A.

---

[10] Although this phenomenon has consistently occurred for group, ring and algebra homomorphisms, it is not true that there is such an equivalence of the existence of an inverse morphism in the category and the bijectivity of the underlying function in arbitrary concrete categories.

Solution. We have that

$$(x^5 + I) - (1 + I) = (x^5 - 1) + I = I = 0 + I$$

since $x^5 - 1 \in I$. Hence, in A, $x^5 + I = 1 + I$. The other basis elements $x^i + I$ for $0 \leq i \leq 4$ remain linearly independent, so a basis is

$$\mathcal{B}' = \{1 + I, x + I, x^2 + I, x^3 + I, x^4 + I\}.$$

We will see this algebra again in another guise later.

1.6 Quivers

While you should have met groups and rings before reading this book, it is less likely you will have encountered quivers. Quivers themselves are not at all complicated, as we will see shortly, but their representation theory is; indeed, some very small quivers have (in a precise mathematical sense) infinitely more complicated representation theory than any finite group. This might sound scary but is actually a positive: we can explore more advanced notions of representation theory in small and concrete examples.

A quiver is just a directed graph, but (as so often happens in mathematics) it comes with some special terminology, some of which does not match conventional usage in graph theory. The definitions below are mainly those commonly used in representation theory, with a little bias towards the author's own preferences in the context of this book as a whole.

Definition 1.6.1. A *quiver* $\mathcal{Q}$ consists of

- a set of *vertices* $\text{Vert}(\mathcal{Q})$ and

- for each pair of vertices $v, w \in \text{Vert}(\mathcal{Q})$ a collection of *arrows* $\mathcal{Q}(v, w)$.

Remark 1.6.2 (If you have never seen the definition of a category, this remark is best left on a first reading and returned to after Chapter 2.).

Comparing with the definition of a category (2.1.1), we will shortly be able to see that a quiver could be called a "pre-category". The vertices correspond to the objects and the arrows to morphisms, but in a quiver we do not ask for any further properties to hold. There is a subtle difference in that we ask for a quiver to have a set of vertices, rather than a collection, but mostly this is just to skirt around set-theoretic issues; in practice, our quiver will be *small* in the category theory sense, i.e. the arrow collections will also be sets.

We say that a quiver $\mathcal{Q}$ is *finite* if $\text{Vert}(\mathcal{Q})$ is finite and $\mathcal{Q}(v, w)$ is finite for every $v, w \in \text{Vert}(\mathcal{Q})$. An element of $\mathcal{Q}(v, v)$ is called a *loop* at v; note that $\mathcal{Q}(v, v)$ can be empty (unlike in a category). If $\alpha \in \mathcal{Q}(v, w)$, the vertex v is called the *tail* $t(\alpha)$ of α and w its *head* $h(\alpha)$. We depict this visually as $v \xrightarrow{\alpha} w$.

If $v \in \mathrm{Vert}(\mathcal{Q})$ has the property that for all $w \in \mathrm{Vert}(\mathcal{Q})$, $\mathcal{Q}(v,w) = \emptyset$ we say that v is a *sink* (as v has no arrows leaving it, only possibly entering). Conversely if $w \in \mathrm{Vert}(\mathcal{Q})$ is such that for all $v \in \mathrm{Vert}(\mathcal{Q})$, $\mathcal{Q}(v,w) = \emptyset$ then we say that w is a *source* (no arrows enter w, they can only possibly leave). A vertex which is both a source and a sink is called *isolated*. Note that if v has a loop, it is neither a source nor a sink and is not isolated, even if there are no other arrows to or from v.

Examples 1.6.3. In what follows, where we can, we draw quivers in the natural way, rather than formally specifying $\mathrm{Vert}(\mathcal{Q})$ and $\mathcal{Q}(v,w)$. If $\mathrm{Vert}(\mathcal{Q})$ is finite of size n, we usually use $\{1, \ldots, n\}$ as the labelling set for the vertices; we will tend to use Greek letters to name arrows.

(a) the quiver A_1 with one vertex and no arrows, 1

(b) the quiver L_1 with one vertex and one loop, 1 ⟳

(c) the quiver A_2 with two vertices and one arrow, $1 \longrightarrow 2$

(d) the *Kronecker quiver* K_2, $1 \rightrightarrows 2$

(e) the *3-subspace quiver*, Sub_3,

$$
\begin{array}{ccc}
 & 1 & \\
 & \downarrow & \\
2 \longrightarrow & 4 & \longleftarrow 3
\end{array}
$$

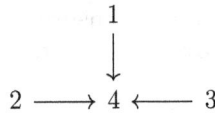

(f) the *oriented cycle with 4 vertices* C_4,

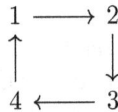

$$
\begin{array}{ccc}
1 & \longrightarrow & 2 \\
\uparrow & & \downarrow \\
4 & \longleftarrow & 3
\end{array}
$$

(g) a quiver chosen to have all of the above features,

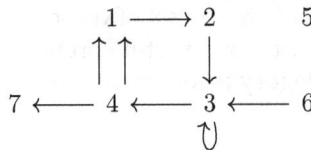

$$
\begin{array}{ccccc}
 & 1 & \longrightarrow & 2 & & 5 \\
 & \uparrow\uparrow & & \downarrow & & \\
7 \longleftarrow & 4 & \longleftarrow & 3 & \longleftarrow & 6 \\
 & & & \circlearrowleft & &
\end{array}
$$

in which 5 is isolated, 6 is a source and 7 is a sink.

SageMath can display and manipulate quivers, under the name DiGraph, for "directed graph". For example, the following code displays Sub$_3$ from (e) above:

```
1  Q = DiGraph([[1,2,3,4],[(1,4),(2,4),(3,4)]],format
       ='vertices_and_edges')
2  Q
```

This approach first gives SageMath a list of vertices, [1,2,3,4], then a list of directed edges (so (1,4) says put a directed edge from 1 to 4) and a format= key to tell SageMath this is the format we're using. There are several different permitted formats for digraphs (quivers) in SageMath and the system tries to work out which you meant: adding the key skips this interpretation.

An alternative syntax for the same graph is

```
1  Q = DiGraph({1:[4],2:[4],3:[4]})
2  Q
```

Here, for each vertex we say which are the arrows out of that vertex and label the arrow. Sometimes it is necessary for the arrows to be given names, which we can do via a tweak to the latter format (note the changes in use of braces { } versus parentheses ( ) or square brackets [ ]).

```
1  Q = DiGraph({1:{4:'e14'},2:{4:'e24'},3:{4:'e34'}})
2  Q
```

We have a notion of subobject for quivers:

Definition 1.6.4. Let Q be a quiver. We say that a quiver Q' is a subquiver of Q if Vert$(Q') \subseteq$ Vert(Q) and $Q'(v,w) \subseteq Q(v,w)$ for all $v,w \in$ Vert(Q').

That is, a subquiver of Q consists of some of the vertices and some of the arrows of Q.

Remark 1.6.5. Note that as we do not have a notion of homomorphism of quivers, we have to ask that the set of vertices and collections of arrows defining Q' are actually subsets and sub-collections of those of Q. It would, of course, be better to have a notion of morphism so that we could talk more generally about Q' being isomorphic to a subquiver of Q. We leave it as an exercise to think about how such a set of definitions would be constructed.

When we have a graph, especially a directed graph, it is natural to consider paths.

Definition 1.6.6. Let $\mathcal{Q}$ be a quiver. A finite *path* in $\mathcal{Q}$ is a sequence

$$(v_0, \alpha_0, v_1, \alpha_1, \ldots, v_{n-1}, \alpha_{n-1}, v_n)$$

of vertices v_i and arrows α_i of $\mathcal{Q}$ such that $\alpha_i \in \mathcal{Q}(v_i, v_{i+1})$ for $0 \leq i \leq n-1$ (i.e. $t(\alpha_i) = v_i$ and $h(\alpha_i) = v_{i+1}$). We say that such a sequence is a path of length n from v_0 to v_n.

For any vertex $v \in \mathrm{Vert}(\mathcal{Q})$, we have the *trivial path* $e_v \overset{\mathrm{def}}{=} (v, v)$ of length 0.

For $v, w \in \mathrm{Vert}(\mathcal{Q})$, let $\mathcal{P}(v, w)$ denote the collection of finite paths from v to w.

Our definition does not allow for infinite paths, as handling these (when they exist) requires significant extra technical complication and formalism, so from this point, we will use "path" to mean "finite path" (unless otherwise stated). Also, we impose no restrictions on the vertices and arrows beyond those stated, so paths may visit the same vertex or use the same arrow multiple times.

We can now formalize the phrase "oriented cycle" used above.

Definition 1.6.7. Let $\mathcal{Q}$ be a quiver. An *oriented cycle* in $\mathcal{Q}$ is a finite path $(v_0, \alpha_0, v_1, \alpha_1, \ldots, v_{n-1}, \alpha_{n-1}, v_n)$ of length $n \geq 1$ such that $v_0 = v_n$.

Note that paths of length 0 are not considered to be oriented cycles but we do allow oriented cycles of length 1 (when they are loops).

Definition 1.6.8. We say a quiver $\mathcal{Q}$ is *acyclic* if it has no oriented cycles.

For example, the Kronecker quiver is acyclic but the quiver C_4 and that in Examples 1.6.3(g) are not.

Continuing the above example,

```
3   Q.is_directed_acyclic()
```

returns True.

We also have a notion of how many 'pieces' a quiver has.

Definition 1.6.9. Let $\mathcal{Q}$ be a quiver. We say that $\mathcal{Q}$ is *connected* if for every pair of vertices $v, w \in \mathrm{Vert}(\mathcal{Q})$, there exists a sequence

$$(v = v_0, p_0, v_1, p_1, \ldots, v_{n-1}, p_{n-1}, v_n = w)$$

of vertices v_i and paths p_i in $\mathcal{Q}$ such that either p_i is a path from v_i to v_{i+1} or p_i is a path from v_{i+1} to v_i.

It is not hard to see that every quiver can be regarded as a disjoint union of connected subquivers. With a little thought, one also sees that this definition is equivalent to saying that the *underlying undirected graph* of the quiver is connected in the usual graph-theoretic sense, where the underlying undirected graph is obtained by replacing each arrow by an (unoriented) edge between the corresponding vertices.

Also,

4   `Q.is_connected()`

returns True.

1.E   Exercises

Remark 1.E.1. In the Exercises sections, we will provide more questions for you to investigate, to support your learning. This is in addition to the Problems already included in the text (which we sometimes repeat in these sections, if we are particularly keen to recommend them to you). By comparison with the Challenges, they should be accessible to you with a bit of thinking, although some will be more difficult and may be marked "Hard" if appropriate.

We encourage you to use SageMath or another computer algebra system if you find that helpful. You may need (or want) to use functionality beyond what is described in the main text (although it is an interesting challenge to see what you can do yourself, versus always reaching for inbuilt functions) and there is often more than one solution that works.

We have chosen not to give a large number of questions on groups or rings in this chapter, because that is not the focus of this book. Rather, here we have concentrated on things you will particularly need for representation theory and/or that are likely to be new to you.

Exercise 1.1. By hand(!), find the eigenvectors and eigenvalues of the following (complex) matrices. Say whether or not they are diagonalisable and if they are, diagonalize them.

(a) $\begin{pmatrix} 3 & -2 \\ 2 & -2 \end{pmatrix}$

(b) $\begin{pmatrix} 2 & 1 & 0 \\ 0 & 2 & 0 \\ 0 & 0 & 0 \end{pmatrix}$

(c) $\begin{pmatrix} 3 & -2 & 1 \\ 2 & -2 & 2 \\ 1 & 2 & -3 \end{pmatrix}$

Check your answers using your preferred computer algebra system.

Exercise 1.2. Are the following pairs of complex matrices similar? Briefly say why or why not.

(a) $\begin{pmatrix} 3 & -2 \\ 2 & -2 \end{pmatrix}, \begin{pmatrix} 2 & 0 \\ 0 & -1 \end{pmatrix}$

(b) $\begin{pmatrix} 3 & -2 \\ 2 & -2 \end{pmatrix}, \begin{pmatrix} 3 & 0 \\ 2 & -1 \end{pmatrix}$

(c) $\begin{pmatrix} 2 & 0 & 5 \\ 0 & 0 & 0 \\ 0 & 0 & 2 \end{pmatrix}, \begin{pmatrix} 2 & 1 & 0 \\ 0 & 2 & 0 \\ 0 & 0 & 0 \end{pmatrix}$

Exercise 1.3. Check that the construction of the quotient vector space V/W in Section 1.4.1 does indeed yield a vector space.

Exercise 1.4. Let V be a 2-dimensional vector space over $\mathbb{K}$ with basis $\mathcal{B}_V = \{v_1, v_2\}$ and let W be a 3-dimensional vector space over $\mathbb{K}$ with basis $\mathcal{B}_W = \{w_1, w_2, w_3\}$. Write down bases for $V \otimes W$ and $W \otimes V$ and hence find their dimensions.

Exercise 1.5. Let V be a vector space over $\mathbb{K}$ and $V \otimes V$ its tensor product with itself. Show that there is a subspace $1_{\mathbb{K}} \otimes V$ of $V \otimes V$ spanned by all elements of the form $1_{\mathbb{K}} \otimes v$ for $v \in V$. Furthermore, show that the quotient $V \otimes V / 1_{\mathbb{K}} \otimes V$ is isomorphic to V.

[*You may wish to assume V is finite-dimensional first, so that you can choose a finite basis and see what is happening using this, before trying to write a proof that works in general.*]

Exercise 1.6. Let $\mathbb{K}$ be a field. Find an algebra homomorphism from $\mathbb{K}$ to $M_n(\mathbb{K})$.

[*You might prefer to start with the special case $\mathbb{K} = \mathbb{R}$ and $n = 2$.*]

Exercise 1.7 (Similar to Problem 7). Let $A = \mathbb{C}[x]/I$ be the algebra given by the quotient of the polynomial algebra in one variable over $\mathbb{C}$, $\mathbb{C}[x]$, by the ideal $I = \langle x^6 - 1 \rangle$ generated by $x^6 - 1$. Find a basis for A and write down the addition and multiplication tables for A with respect to this basis. That is, for each $b, c \in \mathcal{B}$ in your basis for A, give the addition table having $b + c$ in its (b, c)-entry and the multiplication table with bc in its (b, c)-entry.

Exercise 1.8. Write down some examples of quivers with and without

(a) loops,

(b) sources,

(c) sinks,

(d) isolated vertices or

(e) oriented cycles.

Try to think of examples that are different from those in Examples 1.6.3. Are any of your examples subquivers of any of the others? Are your examples connected or disconnected?

Chapter 2

Categories

In this chapter, we will introduce some new language that will help us talk about representation theory. Categories are, as I hope you will see, a very natural way to express relationships among collections of algebraic objects with a particular structure, in a way that respects the natural functions between them. We will not need very much actual category *theory* but using the terminology will allow us to make better, more precise statements.

2.1 Categories

We start with the definition.

Definition 2.1.1. A *category* $\mathcal{C}$ consists of

- a collection[1] of *objects* $\mathrm{Obj}(\mathcal{C})$ and

- for each pair of objects $x, y \in \mathrm{Obj}(\mathcal{C})$ a collection of *morphisms* $\mathcal{C}(x, y)$

such that

(a) for each pair of morphisms $f \in \mathcal{C}(x, y)$ and $g \in \mathcal{C}(y, z)$, there exists a *composite* morphism $g \circ f \in \mathcal{C}(x, z)$ such that *composition is associative*:

$$h \circ (g \circ f) = (h \circ g) \circ f$$

for all $f \in \mathcal{C}(w, x)$, $g \in \mathcal{C}(x, y)$ and $h \in \mathcal{C}(y, z)$; and

(b) for each $x \in \mathrm{Obj}(\mathcal{C})$, there exists an *identity morphism* $\mathrm{id}_x \in \mathcal{C}(x, x)$ such that *the left and right unit laws hold*:

$$\mathrm{id}_y \circ f = f = f \circ \mathrm{id}_x$$

for all x, y and $f \in \mathcal{C}(x, y)$.

As you might expect, there are lots of examples.

---

[1]Here and elsewhere, we will say "collection" due to what mathematicians call "size issues". These occur because the collection of all things we might be interested in might not form a set, depending on our logical foundations. You might recall that issues such as Russell's paradox can occur if care is not taken. We will not dwell on this but will signal that care is needed, by using "collection" instead of "set".

©2025 Jan E. Grabowski, CC BY-NC 4.0 https://doi.org/10.11647/OBP.0492.02

Examples 2.1.2.

(a) There is a category with one object $*$ and one morphism id_*.

(b) Let $v \xrightarrow{a} w$ be the directed graph Q with two vertices v, w and one edge a. Then there is a category $\mathcal{P}(Q)$ with two objects, v and w, and $\mathcal{P}(v, v) = \{\mathrm{id}_v\}$, $\mathcal{P}(v, w) = \{a\}$ and $\mathcal{P}(w, w) = \{\mathrm{id}_w\}$.

(c) Let G be a group. Let $\mathcal{G}$ be the category with one object $*$ and $\mathcal{G}(*, *) = G$. Note that $e_G = \mathrm{id}_*$.

But, I hear you say, these examples are all a bit ... unnatural. Yes: this makes the point that the definition of a category includes a huge range of structures.

Fortunately, there are more familiar examples; in the sections following this, we will recap many of the definitions below, should you need them.

(a) Let **Set** be the category with objects $\mathrm{Obj}(\textbf{Set})$ being the collection of all sets and with morphisms $\textbf{Set}(X, Y) = \mathcal{F}(X, Y)$, that is, the collection of morphisms between two sets X and Y is the collection of all functions from X to Y. Then you have probably known for a long time that composition of functions exists, is associative and that we have identity functions.

(b) Let **Grp** be the category with objects $\mathrm{Obj}(\textbf{Grp})$ being the collection of all groups and with morphisms $\textbf{Grp}(G, H) = \mathrm{Hom}(G, H)$ the collection of all group homomorphisms from G to H. Composition and its associativity are inherited from **Set** and the identity function is a group homomorphism.

(c) Let **Rng** be the category with objects $\mathrm{Obj}(\textbf{Rng})$ being the collection of all (not necessarily unital) rings and with morphisms $\textbf{Rng}(R, S) = \mathrm{Hom}(R, S)$ the collection of all (not necessarily unital) ring homomorphisms from R to S. Composition and its associativity are inherited from **Set** and the identity function is a ring homomorphism.

(d) Let **Ring** be the category with objects $\mathrm{Obj}(\textbf{Ring})$ being the collection of all unital rings and with morphisms $\textbf{Ring}(R, S) = \mathrm{Hom}_u(R, S)$ the collection of all unital ring homomorphisms from R to S. Composition and its associativity are inherited from **Set** and the identity function is a ring homomorphism.

(e) Let $\mathbb{K}$ be a field and let $\textbf{Vect}_{\mathbb{K}}$ be the category with objects $\mathrm{Obj}(\textbf{Vect}_{\mathbb{K}})$ being the collection of all $\mathbb{K}$-vector spaces and with morphisms $\textbf{Vect}_{\mathbb{K}}(V, W) = \mathrm{Hom}_{\mathbb{K}}(V, W)$ the collection of all $\mathbb{K}$-linear maps from V to W. Composition and its associativity are inherited from **Set** and the identity function is a $\mathbb{K}$-linear map.

(f) Let **Top** be the category with objects Obj(**Top**) being the collection of all topological spaces and with morphisms $\textbf{Top}(X, Y) = C^0(X, Y)$ the collection of all continuous functions from X to Y.

(Don't worry if you don't know this example: it won't appear again in this course, and this is why it was included, i.e. to illustrate that there are examples that are natural but not algebraic. Topological spaces are not usually thought of as *algebraic* structures, but they are sets with some data—the data of a topology—and continuous maps are by definition functions preserving the topology, so they do broadly fit in here.)

Most of our categories will be like these, with morphisms being some structure-preserving map, but not all are. The following generalizes (b) above.

Definition 2.1.3. Let $\mathcal{Q}$ be a quiver. The *path category* $\mathcal{P}(\mathcal{Q})$ of $\mathcal{Q}$ is the category with objects Vert($\mathcal{Q}$) and morphisms $\mathcal{P}(\mathcal{Q})(v, w) = \mathcal{P}(v, w)$ for $v, w \in$ Vert($\mathcal{Q}$).

That is, the vertices of $\mathcal{Q}$ are the objects and the morphisms are, as the name might suggest, all (finite) paths.

This is a category. Composition of morphisms is given by concatenation of paths (when this is possible), i.e. for

$$p = (v_0, \alpha_0, v_1, \alpha_1, \ldots, v_{n-1}, \alpha_{n-1}, v_r) \in \mathcal{P}(\mathcal{Q})(v_0, v_r)$$

and

$$p' = (v'_0, \alpha'_0, v'_1, \alpha'_1, \ldots, v'_{s-1}\alpha'_{s-1}, v'_s) \in \mathcal{P}(\mathcal{Q})(v'_0, v'_s),$$

$p' \circ p$ is defined if $v_r = v'_0$ and then

$$p' \circ p = (v_0, \alpha_0, v_1, \alpha_1, \ldots, v_{n-1}, \alpha_{n-1}, v_r = v'_0, \alpha'_0, v'_1, \alpha'_1, \ldots, v'_{s-1}, \alpha'_{s-1}, v'_s).$$

The identity morphism at $v \in$ Vert($\mathcal{Q}$) is the trivial path e_v at v. One may readily check the required properties (exercise).

Going forward, our usual convention for morphism sets in general categories will be to write $\text{Hom}_{\mathcal{C}}(X, Y)$ for morphisms from X to Y in the category $\mathcal{C}$, to remind us to think of homomorphisms.

When $X = Y$, we have a special name for the resulting morphisms; namely, we write $\text{End}_{\mathcal{C}}(X) \overset{\text{def}}{=} \text{Hom}_{\mathcal{C}}(X, X)$. A morphism $f\colon X \to X$ is called an *endomorphism*.

Some other important types of morphism, with names that should by now be familiar, are as follows.

Definition 2.1.4. Let $\mathcal{C}$ be a category. A morphism $f\colon x \to y$ is an *isomorphism* if it is invertible, i.e. there exists $g\colon y \to x$ such that $g \circ f = \text{id}_x$ and $f \circ g = \text{id}_y$.

Given $x, y \in \mathcal{C}$, if there exists an isomorphism $f: x \to y$ between them, we say x and y are *isomorphic* and write $x \cong y$. An isomorphism $f: x \to x$ is called an *automorphism* of x and we write $\mathrm{Aut}_\mathcal{C}(x)$ for the collection of automorphisms of $x \in \mathcal{C}$.

Lemma 2.1.5. *For any category $\mathcal{C}$ and object $x \in \mathcal{C}$, $\mathrm{Aut}_\mathcal{C}(x)$ is a group.*

Proof. The composition of two automorphisms is well-defined and again an automorphism: if $f_1, f_2 \in \mathrm{Aut}_\mathcal{C}(x)$ have inverses g_1, g_2 respectively, then $(g_2 \circ g_1) \circ (f_1 \circ f_2) = \mathrm{id}_x$ and similarly in the other order. The identity morphism is an automorphism $(\mathrm{id}_x \circ \mathrm{id}_x = \mathrm{id}_x)$ and automorphisms have inverses by definition. $\qquad\square$

Perhaps surprisingly, SageMath is aware of categories as an organizational tool for its algorithms. So one can ask things like

```
1  R.<x> = PolynomialRing(QQ)
2  R.categories()
```

and find out the rather impressive list of categories to which R is known to SageMath to belong. Much of SageMath's functionality in this area is aimed at a higher level than we are working at, so this comment is mainly just for interest.

2.2 Functors

Just as sets, groups, rings, vector spaces etc. have structure-preserving maps between them, so do categories. Indeed, it was claimed by Mac Lane, one of the founders of category theory, that being able to define and study these was the whole point. Indeed, just as bijections or group or ring isomorphisms or invertible linear maps encode for us when two sets, groups, rings or vector spaces are "the same", so certain maps of categories will do this too.

Definition 2.2.1. Let $\mathcal{C}$ and $\mathcal{D}$ be categories. A *functor* $F: \mathcal{C} \to \mathcal{D}$ is a map that sends every $x \in \mathcal{C}$ to an object $Fx \in \mathcal{D}$ and every morphism $f \in \mathcal{C}(x, y)$ to a morphism $Ff \in \mathcal{D}(Fx, Fy)$ such that

(a) *F preserves composition:* $F(g \circ f) = Fg \circ Ff$ for all composable f, g and

(b) *F preserves identity morphisms:* $F\mathrm{id}_x = \mathrm{id}_{Fx}$ for all $x \in \mathrm{Obj}(\mathcal{C})$.

It should be clear by comparing with the definition of a category that this is the[2] natural definition that preserves the structure within a category.

The preservation conditions in the definition of a functor could be gathered together into a slicker statement: that F preserves all commuting diagrams. We will use commuting diagrams regularly in our definitions, so let us expand on this a little.

Informal definition 2.2.2. A *diagram* in a category $\mathcal{C}$ is a collection of objects and morphisms between them, some of which may be composable. We say that a diagram *commutes*, or is a *commuting diagram*, if the compositions of the morphisms along any two paths with the same start and end objects are equal.

For example, consider

$$
\begin{array}{ccc}
x & \xrightarrow{\ f\ } & y \\
 & \searrow{\scriptstyle g\circ f} & \downarrow{\scriptstyle g} \\
 & & z
\end{array}
$$

Here, g and f are composable and the composition is also in the diagram. We see that the two (directed) paths from x to z indeed have equal compositions along them. These are the only two non-trivial paths so the diagram commutes. Note that, for example, the morphisms g and $g \circ f$ are not composable; this is permitted and just means that there are no further conditions to check involving these.

The more formal definition would involve functors from the path category of a quiver to $\mathcal{C}$, but we will not need this level of precision.

Then the collection of all categories together with functors between them forms a category. Unsurprisingly, we are running headlong into logical and set-theoretical issues here again, but we can avoid some of this, as follows.

Definition 2.2.3. A category $\mathcal{C}$ is called *locally small* if for all $x, y \in \mathrm{Obj}(\mathcal{C})$, $\mathcal{C}(x, y)$ is a set. A category $\mathcal{C}$ is called *small* if $\mathrm{Obj}(\mathcal{C})$ is a set and $\mathcal{C}$ is locally small.

Problem 8 (Hard and requires some set theory!). Which of the above examples of categories are small? locally small?

Definition 2.2.4. The category **Cat** with objects all small categories and morphisms **Cat**$(\mathcal{C}, \mathcal{D})$ all functors from $\mathcal{C}$ to $\mathcal{D}$ is a category.

An important class of functors, especially given our examples above, is that of *forgetful functors*. There is not a precise definition but the examples

---

[2]In fact, it is one of two such, though arguably the other looks less natural at first sight. Namely, what we have defined is called a *covariant* functor; its counterpart, a *contravariant* functor, satisfies $F(g \circ f) = Ff \circ Fg$.

give the idea: when an algebraic structure consists of a set with additional data, there is a functor to the category **Set** sending every object to itself but *forgetting* the extra structure. We also send morphisms, which we typically take to be functions that preserve the additional structure, to themselves, forgetting that they have extra properties. We don't have to forget *all* the structure, either. In this way we have forgetful functors as follows:

- $\mathcal{F}\colon \mathbf{Ring} \to \mathbf{Set}$

- $\mathcal{F}\colon \mathbf{Rng} \to \mathbf{Set}$

- $\mathcal{F}\colon \mathbf{Grp} \to \mathbf{Set}$

- $\mathcal{F}\colon \mathbf{Vect}_{\mathbb{K}} \to \mathbf{Set}$

- $\mathcal{F}\colon \mathbf{Ring} \to \mathbf{Rng}$

- $\mathcal{F}\colon \mathbf{Ring} \to \mathbf{Grp}$

- $\mathcal{F}\colon \mathbf{Rng} \to \mathbf{Grp}$

- $\mathcal{F}\colon \mathbf{Top} \to \mathbf{Set}$

Most functors actually "do" something and we will see more non-trivial examples later.

The first four of these and the last express that the categories in the domain of $\mathcal{F}$ (i.e. our more familiar examples of categories) are what is called *concrete*. The existence of these forgetful functors to **Set** is the precise way to say that these categories consist of sets and functions with extra structure.

Especially in these examples, but also when we are generally feeling lazy, we will write $x \in \mathcal{C}$ to mean "x is an object of $\mathcal{C}$" (i.e. $x \in \mathrm{Obj}(\mathcal{C})$). So for example we might say "let $V \in \mathbf{Vect}_{\mathbb{K}}$" as a shorthand for "let V be a $\mathbb{K}$-vector space".

2.3   Natural transformations and equivalences

We will continue our whirlwind tour of "elementary category theory", if there is such a thing, by addressing the issue mentioned above of when two categories should be considered to be "the same". Perhaps surprisingly,[3] there are several levels of sameness, and the most direct analogue of isomorphism is not the right one.

The first idea we might have is to say that two categories $\mathcal{C}$, $\mathcal{D}$ are the same if there is an isomorphism $F\colon \mathcal{C} \to \mathcal{D}$ in **Cat** between them. That is, if there are two functors $F\colon \mathcal{C} \to \mathcal{D}$ and $G\colon \mathcal{D} \to \mathcal{C}$ such that $G \circ F = \mathrm{id}_{\mathcal{C}}$ and $F \circ G = \mathrm{id}_{\mathcal{D}}$, where $\mathrm{id}_{\mathcal{C}}$ is the identity functor on $\mathcal{C}$ (sending every object to itself and every morphism to itself).

---

[3]Until one has spent a bit more time moving around the hierarchy of categories.

However this is too strict a notion and isomorphisms of categories are very rare. Instead, we should relax the conditions a little and say that $G \circ F$ and $F \circ G$ should be *almost* the identity functors. To do this definition properly requires the notion of a *natural transformation* of functors. We will also show that this is equivalent to verifying some properties that are both easier to check and more like the "injective and surjective" condition we are used to in **Set**.

Definition 2.3.1. Let C and D be categories and let $F \colon C \to D$ and $G \colon C \to D$ be functors. A *natural transformation* $\alpha \colon F \Rightarrow G$ is an assignment to every object $x \in C$ a morphism $\alpha_x \in D(Fx, Gx)$, called the *component* of α at x, such that for any $f \in C(x, y)$ the following diagram in D commutes:

$$
\begin{array}{ccc}
Fx & \xrightarrow{\ Ff\ } & Fy \\
{\scriptstyle \alpha_x}\downarrow & & \downarrow{\scriptstyle \alpha_y} \\
Gx & \xrightarrow[\ Gf\ ]{} & Gy
\end{array}
$$

Note that natural transformations can be composed: given $\alpha \colon F \Rightarrow G$ and $\beta \colon G \Rightarrow H$ (for $H \colon C \to D$ also), define $\beta \circ \alpha \colon F \Rightarrow H$ to be the natural transformation with components $\beta_x \circ \alpha_x$. That this is a natural transformation is shown by stacking the commutative square on top of its analogue for G and H. This is the beginning of showing that **Cat** is a *2-category*.

Definition 2.3.2. A natural transformation $\alpha \colon F \Rightarrow G$ is called a *natural isomorphism* if every component $\alpha_x \colon Fx \to Gx$, $x \in C$, is an isomorphism. We write $F \cong G$ if F and G are (naturally)[4] isomorphic.

Now, we may give the definition suggested above.

Definition 2.3.3. Let C and D be categories. We say C and D are *equivalent* if there exist functors $F \colon C \to D$ and $G \colon D \to C$ such that $G \circ F \cong \mathrm{id}_C$ and $G \circ F \cong \mathrm{id}_D$.

Definition 2.3.4. Let C and D be (small) categories and $F \colon C \to D$ a functor. For $x, y \in C$ denote by $F_{xy} \colon C(x, y) \to D(Fx, Fy)$ the function defined by $F_{xy}(f) = Ff$.

(a) We say that F is *faithful* if F_{xy} is injective for all $x, y \in C$.

(b) We say that F is *full* if F_{xy} is surjective for all $x, y \in C$.

(c) We say that F is *fully faithful* if F_{xy} is bijective for all $x, y \in C$.

(d) We say that F is *essentially surjective* if for every object $d \in D$ there exists $c \in C$ such that $Fc \cong d$.

---

[4] We usually drop the adjective "naturally" as we will not want to consider unnatural isomorphisms.

Proposition 2.3.5. *Let C and D be (small) categories and $F: C \to D$ a functor. Then F is an equivalence of categories if and only if F is fully faithful and essentially surjective.*

Proof. See [ML, Theorem 1, §IV.4]. □

We will see several examples later.

2.4   Universal properties

When we considered the First Isomorphism Theorems for group and rings and when we defined tensor products, we saw that it can be very productive to give certain definitions and constructions via *universal properties*. The name is intended to convey that for some given inputs, there is a distinguished object having the desired properties with respect to those inputs.

In general, the advantages of this approach are that we get what we want uniquely up to unique isomorphism (which is as unique as it is reasonable to expect, in category theory), along with an appropriate map. On the downside, we have to do some work to prove existence. However, once we have constructed one example satisfying the required criteria—which we call a *model*—we know that any other is isomorphic to this.

There is a formal theory of universal properties, as explained in e.g. [Rie], but rather than take this approach, we will instead look at a small number of key examples, some of which you may have met before (although perhaps not expressed in this way).

2.4.1   Kernels

Consider a ring homomorphism $f: R \to S$ for (not necessarily unital) rings R and S; what we are about to say applies almost equally well to groups, vector spaces and other settings, but for relative concreteness let us pick rings.

The kernel of f is usually defined in terms of elements to be

$$\mathrm{Ker}\, f = \{r \in R \mid f(r) = 0_S\}.$$

However, a more categorical definition—which we could apply in any category having some basic properties—would go as follows. Here, we use the zero map $0: R \to S$, $0(r) = 0_S$ for all $r \in R$, where R and S are any rings.[5]

---

[5]We do not over-complicate the notation and so just write 0 for this map, the domain and codomain being inferred from the context.

Definition 2.4.1. A *kernel* of the morphism $f\colon R \to S$ is a pair (K, k) with $K \in \textbf{Rng}$ and $k\colon K \to R$ such that

- $f \circ k = 0$ and

$$K \xrightarrow{\; f \circ k = 0 \;} S$$

$$\begin{array}{ccc} K & \xrightarrow{f\circ k=0} & S \\ {\scriptstyle k}\downarrow & \nearrow_{f} & \\ R & & \end{array}$$

- for any $K' \in \textbf{Rng}$ and $k'\colon K' \to R$ such that $f \circ k' = 0$, there is a unique morphism $u\colon K' \to K$ such that $k \circ u = k'$.

It then follows that kernels are unique up to unique isomorphism, as follows. Given two kernels (K, k), (K', k'), we obtain two unique maps $u\colon K' \to K$ and $u'\colon K \to K'$ from the second part of the definition and then consider $u \circ u'\colon K \to K$. From the diagram

we see that the dashed arrow could be given by id_K and also by $u \circ u'$, since

$$k \circ (u \circ u') = (k \circ u) \circ u' = k' \circ u = k.$$

By the uniqueness of the dashed arrow, $u \circ u' = \mathrm{id}_K$. Similarly, $u' \circ u = \mathrm{id}_{K'}$. So u is the unique isomorphism between K and K' such that $k \circ u = k'$ and correspondingly for its inverse u'.

In a kernel (K, k) the morphism k is injective (exercise); equivalently, the kernel of k is $(0, 0)$, the pair consisting of the zero ring and the unique morphism $0\colon 0 \to K$ with $0(0) = 0$.

Furthermore, we can check that kernels exist by exhibiting a model. Let us consider $(\mathrm{Ker}\, f, \iota)$ with $\iota\colon \mathrm{Ker}\, f \to R$ coming from the natural inclusion of Ker f into R, since the former is a subset of the latter. Then for all $r \in$ Ker f, $(f \circ \iota)(r) = f(r) = 0$ by the definition of Ker f.

If K' is a ring and $k' \colon K' \to R$ satisfies $f \circ k' = 0$, then define $u \colon K' \to$ Ker f by $u(s) = k'(s)$ for all $s \in K'$. This is well-defined since $f \circ k' = 0$ implies that $f(u(s)) = f(k'(s)) = 0$ so $u(s) \in$ Ker f. Now for all $s \in K'$, $\iota(u(s)) = u(s) = k'(s)$ so $\iota \circ u = k'$, and hence a candidate morphism exists. It must be unique since if $v \colon K' \to R$ is any other morphism such that $\iota \circ v = k'$ then $\iota(v(s)) = k'(s) = \iota(u(s))$ for all $s \in K'$. But ι is injective so we see that $v(s) = u(s)$ for all $s \in K'$ and hence $v = u$.

So (Ker f, ι) is indeed a kernel of f, and so we usually speak of Ker f as being *the* kernel of f, keeping in the back of our minds that "the" is shorthand for "unique up to unique isomorphism".

Note that the kernel K of a ring homomorphism $f \colon R \to S$ is an ideal, so that K is only a *unital* ring if $K = R$, whence f is the zero map. This is why we say "$K \in$ **Rng**"; indeed, this phenomenon suggests that **Rng** is a nicer category than **Ring**, according to some suitable notion of nicety.

Remark 2.4.2. By a process called (hopefully, for obvious reasons) "reversing all arrows" we obtain a dual definition, that of the *cokernel* (C, p) of a morphism f. By suitably reversing all arrows in the proofs too, we immediately obtain that the cokernel is unique up to unique isomorphism and that the morphism p is surjective.

Also, we know a model for the cokernel of $f \colon A \to B$ in the category of Abelian groups: one can show that the quotient group $B/\operatorname{Im} f$ together with the natural projection $p \colon B \to B/\operatorname{Im} f$ satisfies the definition. This also works for vector spaces.

However there are difficulties for groups or rings, precisely because we know that to form the quotient group or quotient ring one needs a normal subgroup or an ideal, respectively, and the image of f need not be such.

2.4.2   Quotients

Quotient groups, rings and vector spaces have universal properties, the ring version being as follows.

Theorem (Universal property of the quotient ring, 1.3.45). *Let R and S be rings. Let I be an ideal of R and let $\pi \colon R \to R/I$ be the associated quotient homomorphism.*

Then for every ring homomorphism $\varphi \colon R \to S$ such that $I \subseteq$ Ker φ, there exists a unique ring homomorphism $\bar{\varphi} \colon R/I \to S$ such that $\varphi = \bar{\varphi} \circ \pi$.

Furthermore, Ker $\bar{\varphi} = ($Ker $\varphi)/I$ and $\operatorname{Im} \bar{\varphi} = \operatorname{Im} \varphi$.

The corresponding commutative diagram is

$$
\begin{array}{ccc}
R & \xrightarrow{\;\varphi\;} & S \\
\pi \downarrow & \nearrow & \\
R/I & \exists!\bar{\varphi} &
\end{array}
$$

This and its analogues for groups (1.2.85) and vector spaces claim that the usual construction in terms of cosets is a model for a quotient object defined by the relevant universal property. In this case, the property is that ideals (respectively, subgroups and subspaces) contained in the kernel of a morphism yield a unique morphism from the quotient to the codomain.

The case of modules is given in Section 4.2.2, along with the isomorphism theorems, which follow directly from the universal property of quotients. That quotients for groups, rings etc. all satisfy a universal property of the same "shape" explains why their isomorphism theorems all look so similar. In essence, one is just checking that the same proof works in several different categories.

2.4.3 Tensor products

We have seen in Definition 1.4.3 that tensor products of vector spaces are defined via a universal property.

2.4.4 Free constructions

An important class defined by universal properties is that of *free* objects in a given category. So, we have free groups, free Abelian groups, free vector spaces and so on. The intuitive definition of a free object is that, given a set S, one can form the group (vector space, etc.) generated by this set, without imposing any relations (hence, free).

The most familiar case is vector spaces: the free vector space on a set S is the vector space spanned by S. By construction, S becomes a basis for the free vector space. For a field $\mathbb{K}$ and set S we will use the notation $\mathbb{K}[S]$ for the free vector space on S.[6]

Less familiar but still not overly complicated is the idea of a free group, which we discussed briefly in Section 1.2.8. Let us redo that construction slightly more formally.

When constructing a model for the free group on a set, because we need elements to have inverses, we take a set $\bar{S} = \{\bar{s} \mid s \in S\}$ ("$\bar{s}$" is just a name

---

[6]If we stuck rigidly to the notation we will introduce shortly in Definition 2.4.3, we would find ourselves writing $\mathbf{Vect}_{\mathbb{K}}[S]$ rather than $\mathbb{K}[S]$. This is a bit cumbersome, so we will not, but instead just note that some authors prefer $\mathbb{K}S$ for the free vector space to distinguish the former from a free algebra over $\mathbb{K}$ (which is not something we consider in this book).

for an element in the set $\bar{S}$, which is in bijection with S) and consider the group whose elements are *words* (that is, finite strings) in the set $S \cup \bar{S}$ with group operation being concatenation and the identity element the empty word $\emptyset$ (the unique word of length zero), modulo the relation $s\bar{s} = \emptyset = \bar{s}s$.

For example, if $S = \{a, b\}$, the free group on S has elements

$$\{\emptyset, a, b, \bar{a}, \bar{b}, aa, ab, ba, bb, a\bar{b}, \bar{b}a, \bar{a}b, b\bar{a}, aaa, \dots\}$$

and the inverse of $ab\bar{a}$ is $a\bar{b}\bar{a}$.

Once this model is established, it is then more common to write s^{-1} for $\bar{s}$.

Other important examples include polynomial rings (1.3.4): $R[z]$ is the free (commutative) R-algebra on one generator.

Formally, though, freeness is defined by a universal property, as follows. We will stick to the case of free objects on a set in this definition.

Definition 2.4.3. Let $\mathcal{C}$ be a concrete category with forgetful functor $\mathcal{F}: \mathcal{C} \to \mathbf{Set}$ and let $S \in \mathbf{Set}$. A *free object* $\mathcal{C}[S]$ of $\mathcal{C}$ on S is an object $\mathcal{C}[S]$ of $\mathcal{C}$ together with an injective function $i: S \hookrightarrow \mathcal{F}\mathcal{C}[S]$ in $\mathbf{Set}$ such that for all objects $c \in \mathcal{C}$ and functions $f: S \to \mathcal{F}c$, there is a unique morphism $g: \mathcal{C}[S] \to c$ in $\mathcal{C}$ such that $\mathcal{F}g \circ i = f$.

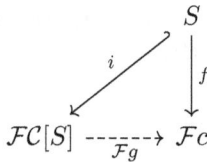

$$
\begin{array}{ccc}
& & S \\
& \overset{i}{\swarrow} & \downarrow f \\
\mathcal{F}\mathcal{C}[S] & \underset{\mathcal{F}g}{\dashrightarrow} & \mathcal{F}c
\end{array}
$$

As usual, free objects are unique up to unique isomorphism.

Much more generally, let $\mathcal{F}: \mathcal{C} \to \mathcal{D}$ be a forgetful functor. Then if a free object $\mathrm{Free}(d)$ exists for all objects $d \in \mathcal{D}$, this assignment can be assembled into a functor $\mathrm{Free}: \mathcal{D} \to \mathcal{C}$ that is left adjoint to $\mathcal{F}$. Adjoints will be discussed in more detail in Section 2.6.

2.5   The Yoneda lemma

The Yoneda lemma, which we will describe in this section, is often considered to be "the first result of category theory". That is, for many people, its statement is the first step beyond basic definitions and checking elementary properties. This philosophy is disputed, though: some argue that— when one has absorbed all the definitions—it is an essential, tautological truth, barely meriting being called a lemma. Indeed, accidentally rediscovering the Yoneda lemma (and only realizing later) has happened to pretty much everyone who works in category theory, including the present author.

We include it mostly out of a sense of completeness, rather than because we will use it heavily. That said, it will make a brief reappearance in Section 6.4.

Let $\mathcal{C}$ be a locally small category; recall that this means that for all $x, y \in \mathrm{Obj}(\mathcal{C})$, $\mathcal{C}(x,y)$ is a set. Then for each $x \in \mathcal{C}$ there is a functor $\mathrm{Hom}_\mathcal{C}(x, -) \colon \mathcal{C} \to \mathbf{Set}$ given on objects by $\mathrm{Hom}_\mathcal{C}(x,-)(y) = \mathrm{Hom}_\mathcal{C}(x,y) \overset{\text{def}}{=} \mathcal{C}(x,y)$. On morphisms, $\mathrm{Hom}_\mathcal{C}(x,-)(f) = \mathrm{Hom}_\mathcal{C}(x,f) = f \circ -$; that is, given $f \colon y \to z$, f is sent to the morphism $g \mapsto f \circ g$ from $\mathrm{Hom}_\mathcal{C}(x,y)$ to $\mathrm{Hom}_\mathcal{C}(x,z)$. This functor is called the (covariant) Hom-functor associated to x. There is a contravariant version too, $\mathrm{Hom}_\mathcal{C}(-, y)$ for each $y \in \mathcal{C}$.

Problem 9. Check that these are indeed functors.[7]

Lemma 2.5.1 (Yoneda). *Let $\mathcal{C}$ be locally small and let $F \colon \mathcal{C} \to \mathbf{Set}$ be a functor. Then for each object $x \in \mathcal{C}$, the natural transformations from $\mathrm{Hom}_\mathcal{C}(x, -)$ to F are in bijection with the set Fx and this bijection is natural.*

One way to understand the Yoneda lemma is as a vast generalization of Cayley's theorem for groups, which we recall and slightly rephrase:

Theorem (Cayley's theorem, 1.2.72). *Let G be a group. There is an injective homomorphism $\lambda \colon G \to \mathrm{Aut}_{\mathbf{Set}}(G)$.*

Here, $\mathrm{Aut}_{\mathbf{Set}}(G)$ (which we originally denoted $\mathrm{Bij}(G)$, before we had categories) is the group of bijections from the underlying set G to itself. Specifically, this theorem is Yoneda for the category $\mathcal{G}$ with one object $*$ and $\mathcal{G}(*, *) = G$ for a group G (Examples 2.1.2(c)). It is left as an exercise to work out the details.

2.6   Adjunction

There is a special sort of equivalence, when the two functors have a particular relationship, as follows.

Definition 2.6.1. Let $\mathcal{C}$ and $\mathcal{D}$ be categories and $F \colon \mathcal{C} \rightleftarrows \mathcal{D} \colon G$ be functors.[8] We say F and G form an *adjoint pair* if for all objects $c \in \mathcal{C}$ and $d \in \mathcal{D}$ there is a bijection

$$\alpha \colon \mathcal{C}(c, Gd) \overset{\cong}{\to} \mathcal{D}(Fc, d)$$

which is natural in c and d.

---

[7] The most difficult part of this exercise is to make sense of the above definition of the functors: diagrams are helpful, as is patience and/or someone willing to listen to you trying to explain it. Indeed this is an instance where the author providing a written solution is not actually very useful, which is something of a relief for me, or perhaps just an excuse.

[8] This notation is shorthand for $F \colon \mathcal{C} \to \mathcal{D}, G \colon \mathcal{D} \to \mathcal{C}$ and emphasizes that we are interested in F and G as a pair. By itself, it makes no claims about $G \circ F$ or $F \circ G$, however.

If F and G form an adjoint pair, we say that F is left adjoint to G and that G is right adjoint to F.

If $\alpha^{-1}(\mathrm{id}_{Fc})\colon c \to GFc$ and $\alpha(\mathrm{id}_{Gd})\colon FGd \to d$ are natural isomorphisms in c and d, we say F and G form an *adjoint equivalence*.

An adjoint equivalence is, in particular, an equivalence.

The reason for the name "adjoint" may be a bit obscure. But we get an insight from recalling that if we have an inner product space, a function f and its adjoint $f^\dagger$ satisfy an equation of the form

$$\langle v, f^\dagger(w)\rangle = \langle f(v), w\rangle$$

and then comparing this to the displayed expression in Definition 2.6.1.

A useful technique one may employ when comparing various categories is to find adjoint pairs (using various general methods and theorems) and then try to show they are actually adjoint equivalences. This is helpful to identify candidate equivalences, just as it is helpful to know some homomorphisms to be able to test for being isomorphisms; without any candidates, it is often not clear what to do.

Examples 2.6.2.

(a) If G is a group, there is an associated Abelian group $G^{\mathrm{ab}} \overset{\text{def}}{=} G/[G,G]$, where $[G,G]$ is the subgroup of G generated by all commutators, that is, all elements of the form $ghg^{-1}h^{-1}$ for $g, h \in G$. The inclusion functor $\mathcal{I}\colon \mathbf{Ab} \to \mathbf{Grp}$ has left adjoint $(-)^{\mathrm{ab}}\colon \mathbf{Grp} \to \mathbf{Ab}$ given on objects by sending G to its Abelianization G^{ab}.

(b) As discussed above, a forgetful functor $\mathcal{F}$ and the functor coming from constructing free objects form an adjoint pair $\mathcal{F}\colon \mathcal{C} \rightleftarrows \mathbf{Set} \colon \mathrm{Free}$.

(c) Products (e.g. the Cartesian product in **Grp**, **Ring** etc.) and kernels are both special cases of a more general categorical construction, that of *limits*. Coproducts (e.g. the direct sum in $\mathbf{Vect}_{\mathbb{K}}$) and cokernels are similarly examples of *colimits*. Functors associated to limits and colimits have adjoints, known as diagonal functors.

(d) As we will see in Section 2.7, tensor products fit into adjoint pairs.

(e) A key example for group representation theory will be proved in Theorem 5.2.8.

2.7   Monoidal categories

2.7.1   Monoidal categories

We briefly introduce the notion of a *monoidal category*, which is a category in which we can form tensor products. This will provide a setting to describe an important example of an adjunction (see 2.7.2) and also lay some groundwork for Hopf algebras (see 6.3). For more on monoidal catgeories and their applications, see the book [EGNO].

First, we need the notion of a *bifunctor*. We say $F\colon \mathcal{C} \times \mathcal{C} \to \mathcal{D}$ is a bifunctor if for all objects $x \in \mathcal{C}$, $F(x,-)\colon \mathcal{C} \to \mathcal{D}$ defined by $F(x,-)(y) = F(x,y)$ and $F(-,x)\colon \mathcal{C} \to \mathcal{D}$ defined by $F(-,x)(y) = F(y,x)$ are respectively covariant and contravariant functors. A key example for later is the Hom bifunctor $\operatorname{Hom}_{\mathcal{C}}(-,-)\colon \mathcal{C} \times \mathcal{C} \to \mathbf{Set}$ on a locally small category $\mathcal{C}$, given by $\operatorname{Hom}_{\mathcal{C}}(-,-)(x,y) = \mathcal{C}(x,y)$.

Definition 2.7.1. A *monoidal category* is a tuple $(\mathcal{C}, \otimes, a, \mathbb{1}, \iota)$ such that

- $\mathcal{C}$ is a category,

- $\otimes\colon \mathcal{C} \times \mathcal{C} \to \mathcal{C}$ is a bifunctor,

- $a\colon (-\otimes -)\otimes - \xrightarrow{\sim} - \otimes (-\otimes -)$ is a natural isomorphism with components

$$a_{xyz}\colon (x \otimes y) \otimes z \xrightarrow{\sim} x \otimes (y \otimes z)$$

- $\mathbb{1} \in \mathcal{C}$ and

- $\iota\colon \mathbb{1} \otimes \mathbb{1} \xrightarrow{\sim} \mathbb{1}$ is an isomorphism

such that

(a) (*the pentagon axiom*) for all $w, x, y, z \in \mathcal{C}$ the following diagram commutes:

(b) (*the unit axiom*) the functors $L_{\mathbb{1}}\colon \mathcal{C} \to \mathcal{C}$, $L_{\mathbb{1}}x = \mathbb{1} \otimes x$ and $R_{\mathbb{1}}\colon \mathcal{C} \to \mathcal{C}$, $R_{\mathbb{1}}x = x \otimes \mathbb{1}$ are autoequivalences[9] of $\mathcal{C}$.

The pair $(\mathbb{1}, \iota)$ is called the unit object of the monoidal category.

Examples 2.7.2.

(a) **Set** is a monoidal category with respect to $\otimes = \times$, the Cartesian product, and $\mathbb{1} = \{*\}$ (a 1-element set).

---

[9]By analogy with automorphisms of algebraic objects, an autoequivalence of a category is an equivalence of the category with itself.

(b) **Vect**$_{\mathbb{K}}$ is a monoidal category with respect to the tensor product of vector spaces $\otimes$ introduced in Section 1.4.1 and $\mathbb{1} = \mathbb{K}$.

(c) The category $\text{Rep}_{\mathbb{K}}(G)$ of linear representations of a group G (as will be discussed in Section 5.2) can be made into a monoidal category, through the tensor product of group homomorphisms ($f \otimes g$ sends $a \otimes b$ to $f(a) \otimes g(b)$).

(d) Below, in Section 4.1, we will introduce the category R-Mod of modules over a commutative unital ring R. This is a monoidal category with $\otimes$ the tensor product of R-modules (defined in exactly the same way as the tensor product of vector spaces) and unit $\mathbb{1} = R$.

We leave as an exercise the task of identifying a suitable map ι in the above examples (in each case, there is a natural/canonical choice).

For completeness, we record here the definition of the special case of the final example for $R = \mathbb{Z}$, that is, when R-Mod $=$ **Ab** (since $\mathbb{Z}$-modules are exactly Abelian groups). This will faciliate defining rings and thence modules and algebras over rings in the way we will wish to do this in Section 4.1.

Definition 2.7.3. Let V and W be Abelian groups. The pair $(X, \otimes)$ is said to be the *tensor product* of V and W if

(a) X is an Abelian group;

(b) $\otimes \colon V \times W \to X$ is a homomorphism of Abelian groups; and

(c) (universal property) for every Abelian group Z and group homomorphism $h \colon V \times W \to Z$, there exists a unique homomorphism $\tilde{h} \colon X \to Z$ such that $h = \tilde{h} \circ \otimes$.

$$V \times W \xrightarrow{\;\otimes\;} X$$
$$\searrow^{h} \quad \downarrow^{\tilde{h}}$$
$$Z$$

By the same argument as in Lemma 1.4.4, X is uniquely determined up to isomorphism. Occasionally we will decorate the symbol $\otimes$ by adding $\mathbb{Z}$ as a subscript to indicate that we are considering this particular tensor product $\otimes_{\mathbb{Z}}$ of Abelian groups. However as this is the bare minimum structure we will consider, we will not use this enhanced notation unless it would be unclear what we meant otherwise. Rather, we will add a subscript R to $\otimes$ if we want to talk about a tensor product $\otimes_R$ over a different commutative unital ring—no subscript is an implied $R = \mathbb{Z}$.

Similarly to the vector space case, the natural model of this tensor product is as a group generated by *elementary tensors* $v \otimes w$ for $v \in V$, $w \in W$. That is, we should take all combinations of elements of the form $v \otimes w$ under the naturally induced group operation. So for $V = \langle g \rangle$ and $W = \langle h \rangle$ two copies of the (generic) infinite cyclic group, we have that $\langle g \rangle \otimes \langle h \rangle$ consists of all elements of the form $g^r \otimes h^s$ with the group operation being $(g^{r_1} \otimes h^{s_1}) * (g^{r_2} \otimes h^{s_2}) = g^{r_1+r_2} \otimes h^{s_1+s_2}$, for $r_i, s_i \in \mathbb{Z}$. This is isomorphic to the free Abelian group of rank 2, i.e. $\langle g, h \mid gh = hg \rangle \cong \mathbb{Z}^2$.

2.7.2 Hom-⊗ adjunction

In certain nice monoidal categories, called *closed* monoidal categories, there is an important relationship between the bifunctors $\operatorname{Hom}(-, -)$ and $\otimes$, which we will now state. We will do this for the particular case of the category R-Mod for R a commutative unital ring, despite not having introduced this category properly; as a category-theoretic tool, it sits better here.

First, we fix $M \in R$-Mod. Then the result says that taking the tensor product $- \otimes M$ with M (on the right) has a natural adjoint, given by taking homomorphisms from M, as per the following result known as Hom-⊗ adjunction (or also ⊗-Hom adjunction, on the grounds that the tensor product is left adjoint to Hom).

Proposition 2.7.4 (Hom-⊗ adjunction). *Let $M \in R$-Mod. There is an adjoint pair*

$$- \otimes M \colon R\text{-Mod} \to R\text{-Mod} \colon \operatorname{Hom}_{R\text{-Mod}}(M, -)$$

so that there exist bijections

$$\alpha \colon \operatorname{Hom}_{R\text{-Mod}}(N, \operatorname{Hom}_{R\text{-Mod}}(M, P)) \xrightarrow{\cong} \operatorname{Hom}_{R\text{-Mod}}(N \otimes M, P)$$

that are natural in N and P.

Proof (sketch). Fix M, N, P as in the statement. For an R-module homomorphism $f \colon N \to \operatorname{Hom}_{R\text{-Mod}}(M, P)$, define $\alpha(f) \colon N \otimes M \to P$ by $\alpha(f)(n \otimes m) = f(n)(m)$. One may check that this is a natural bijection. $\square$

2.E Exercises

Exercise 2.1. Let G and H be groups and let $\mathcal{G}$ and $\mathcal{H}$ be the categories associated to them, as in Examples 2.1.2(c). That is, $\mathcal{G}$ has one object $*$ and $\operatorname{Hom}_{\mathcal{G}}(*, *) = G$ and $\mathcal{H}$ has one object $\bullet$ (we give it a different name to $*$, for clarity) and $\operatorname{Hom}_{\mathcal{H}}(\bullet, \bullet) = H$.

Show that functors $F \colon \mathcal{G} \to \mathcal{H}$ are in one-to-one correspondence with group homomorphisms $f \colon G \to H$.

Exercise 2.2. Consider the quiver A_2 and the 3-subspace quiver Sub_3 from Examples 1.6.3. Show that there is a faithful functor $F\colon \mathcal{P}(A_2) \to \mathcal{P}(\mathrm{Sub}_3)$ between their path categories induced by the subquiver of Sub_3 on the vertices 1 and 4 being a copy of the A_2 quiver.

(More formally, A_2 is isomorphic to the subquiver of Sub_3 on $\{1, 4\}$ and this inclusion induces a faithful functor on the respective path categories.)

Exercise 2.3 (Hard). Think about how you might extend the previous exercise to construct full, fully faithful or essentially surjective functors and hence equivalences between path categories.

Exercise 2.4 (Problem 9). Check that the covarant and contravariant Hom-functors $\mathrm{Hom}_\mathcal{C}(x, -)$ and $\mathrm{Hom}_\mathcal{C}(-, y)$ defined in Section 2.5 are indeed functors.

Chapter 3

Representations

3.1 Representations of groups

3.1.1 Group actions

It is very common to say that groups often arise as symmetries. The subtle but important shift from "what group encodes the symmetries of this object?" to "which objects does this group give symmetries of?" moves us from the *structure theory* of groups to their *representation theory*.

The word "symmetry" needs a little unpacking before we can make much progress—we need some formal definitions in order to state and prove things. Informally, a symmetry of an object should move the points of that object around, there should be a symmetry that leaves the object unchanged and if we have two symmetries, we should be able to apply one and then the other to obtain another symmetry. Very often, we want our symmetry to be reversible; asking for this is what puts us in the domain of groups.

The most common way to encode this formally is in the notion of a group acting on a set, or a group action. We will also do this first but we will then show how this can be transformed into different formulations that are better suited for posing the questions we will want to ask.

Definition 3.1.1. Let G be a group and X a set. A *left action* of G on X is a function $\alpha\colon G \times X \to X$ such that

(GA1) $\alpha(e, x) = x$ for all $x \in X$; and

(GA2) $\alpha(g_1 g_2, x) = \alpha(g_1, \alpha(g_2, x))$ for all $g_1, g_2 \in G, x \in X$.

Problem 10. Write down the definition of a right action.

Problem 11. Write down and prove (using the axioms) a mathematical statement that formalizes the statement "we want a symmetry to be reversible".

For ease of notation, it is common to write $\alpha(g, x)$ as $g \cdot x$. However this can be confusing—the dot $\cdot$ suggests multiplication but this might be unnatural for certain G and X, and the visibility of the "leftness" of the action is reduced.

©2025 Jan E. Grabowski, CC BY-NC 4.0 https://doi.org/10.11647/OBP.0492.03

Instead we will use ▷, i.e. we will write $g \triangleright x$ for $\alpha(g, x)$, so that our axioms become

(GA1′) $e \triangleright x = x$ for all $x \in X$; and

(GA2′) $(g_1 g_2) \triangleright x = g_1 \triangleright (g_2 \triangleright x)$ for all $g_1, g_2 \in G$, $x \in X$.

We will abuse notation a little by naming our action function ▷ too, by writing $\triangleright \colon G \times X \to X, \triangleright(g, x) = g \triangleright x$.

Examples 3.1.2. Important examples of groups acting on sets are:

(a) $\mathrm{GL}_2(\mathbb{R})$ acting on $\mathbb{R}^2$, $\left(\begin{smallmatrix} a & b \\ c & d \end{smallmatrix} \right) \triangleright \left(\begin{smallmatrix} x \\ y \end{smallmatrix} \right) = \left(\begin{smallmatrix} ax+by \\ cx+dy \end{smallmatrix} \right)$;

(b) S_n acting on $\{1, 2, \ldots, n\}$, $\sigma \triangleright r = \sigma(r)$;

(c) G acting on itself via $g \triangleright h = gh$;

(d) G acting on itself via $g \triangleright h = ghg^{-1}$;

(e) G acting on the left cosets of a subgroup H via $g \triangleright kH = gkH$.

Problem 12. Check that these are actions.

Although we will not have so much direct use of the following in this book, it would be remiss not to briefly introduce orbits and stabilizers and the theorem that links them.

Definition 3.1.3. Let $\triangleright \colon G \times X \to X$ be a left action of G on X. The *orbit* of $x \in X$ is the set

$$\mathcal{O}(x) \stackrel{\text{def}}{=} \{g \triangleright x \mid g \in G\}.$$

The *stabilizer* of $x \in X$ is the set

$$\mathrm{Stab}(x) \stackrel{\text{def}}{=} \{g \in G \mid g \triangleright x = x\}.$$

The orbit of an element $x \in X$ is a subset of X and the orbits partition X, i.e. $X = \bigcup_{x \in X} \mathcal{O}(x)$ and $\mathcal{O}(x) \cap \mathcal{O}(y) = \emptyset$ if $y \notin \mathcal{O}(x)$. The stabilizer of x is a subgroup of G.

Problem 13. Prove the claims in the preceding paragraph.

For a group G and subgroup H, we will write $(G : H) = \{gH \mid g \in G\}$ for the set of left cosets of H in G.

Theorem 3.1.4 (Orbit–stabilizer theorem). *Let* $\triangleright \colon G \times X \to X$ *be a left action of* G *on* X. *For* $x \in X$, *there is a bijection*

$$\varphi \colon \mathcal{O}(x) \to (G \colon \mathrm{Stab}(x))$$

defined by $\varphi(g \triangleright x) = g \, \mathrm{Stab}(x)$ *such that*

$$h \triangleright \varphi(y) = \varphi(h \triangleright y)$$

for all $y \in \mathcal{O}(x)$.

Proof. We have

$$g \triangleright x = g' \triangleright x$$
$$\Longleftrightarrow \quad (g')^{-1} \triangleright g \triangleright x = (g')^{-1} \triangleright g' \triangleright x$$
$$\Longleftrightarrow \quad ((g')^{-1}g) \triangleright x = x$$
$$\Longleftrightarrow \quad (g')^{-1}g \in \operatorname{Stab}(x)$$
$$\Longleftrightarrow \quad g \operatorname{Stab}(x) = g' \operatorname{Stab}(x)$$

so that φ is well-defined (two representatives of y in $\mathcal{O}(x)$, $g \triangleright x = g' \triangleright x = y$, map to the same coset) and injective. Since given $g \operatorname{Stab}(x)$, we have $\varphi(g \triangleright x) = g \operatorname{Stab}(x)$, φ is surjective too.

Then

$$h \triangleright \varphi(g \triangleright x) = h \triangleright (g \operatorname{Stab}(x)) = hg \operatorname{Stab}(x) = \varphi(hg \triangleright x) = \varphi(h \triangleright (g \triangleright x)).$$

$\square$

Challenge 14. How would you define a *morphism* between two G-actions? That is, for a fixed G, and given G-actions $\alpha \colon G \times X \to X$ and $\beta \colon G \times Y \to Y$, what would be an appropriate notion of a morphism $f \colon \alpha \to \beta$?

Can you show that the collection $\operatorname{Act}(G, X)$ of (left) actions of G on X with your notion of morphism is a category?

3.1.2 Group representations

Now we make our first reformulation, expressing group actions as group representations.

Recall from Lemma 2.1.5 that the set of automorphisms $\operatorname{Aut}_{\mathcal{C}}(x)$ of an object x in a category $\mathcal{C}$ is a group. For $\mathcal{C} = \textbf{Set}$ and $X \in \textbf{Set}$, $\operatorname{Aut}_{\textbf{Set}}(X) = \operatorname{Bij}(X)$, the group of bijections from X to itself; we will use the former notation, $\operatorname{Aut}_{\textbf{Set}}(X)$, as later we will want to contrast this with groups of automorphisms arising from a different category.

Definition 3.1.5. Let G be a group. A *G-representation* on a set X is a group homomorphism $\rho \colon G \to \operatorname{Aut}_{\textbf{Set}}(X)$.

Definition 3.1.6. Let G be a group and let ρ, σ be G-representations on sets X and Y respectively. A *morphism of G-representations* $f \colon \rho \to \sigma$ is a function $f \colon X \to Y$ such that

$$f(\rho(g)(x)) = \sigma(g)(f(x))$$

for all $g \in G$.

The condition here is best understood as the commuting of the following diagram:

$$
\begin{array}{ccc}
X & \xrightarrow{\rho(g)} & X \\
f\downarrow & & \downarrow f \\
Y & \xrightarrow{\sigma(g)} & Y
\end{array}
$$

Let $\mathrm{Act}(G, X)$ denote the collection of group actions of G on a set X and let $\mathrm{Rep}(G, X)$ denote the collection of G-representations on X.

Challenge 15. Show that $\mathrm{Rep}(G, X)$ with objects G-representations and morphisms as above is a category.

The next proposition says that actions and representations are in one-to-one correspondence. This formally justifies our swapping between the two ideas, or preferring one over the other. We will see similar types of results later, but the methods of proof will differ. Indeed, although a proof is given now for completeness, we recommend skipping it on a first reading, returning to it later if you feel you need to work through the details.

Proposition 3.1.7. *There is a bijection* $\hat{\kappa}\colon \mathrm{Act}(G, X) \to \mathrm{Rep}(G, X)$.

Proof. We first sketch what is needed:

(a) note that $\mathrm{Act}(G, X) \subseteq \mathrm{Hom}_{\mathbf{Set}}(G \times X, X)$;

(b) there is a very general operation κ called "currying"[1] that takes a function $f\colon A \times B \to C$ to a function $\kappa(f)\colon A \to \mathrm{Hom}_{\mathbf{Set}}(B, C)$, defined by $\kappa(f)(a) = f(a, -)$;

(c) the special properties of group actions mean that we can tweak this to replace Hom with Aut;

(d) the resulting functions $\hat{\kappa}(\alpha)$ are group homomorphisms, again because we have a group action, so we can regard $\hat{\kappa}$ as a function taking actions to group representations;

(e) $\hat{\kappa}$ is a bijection.

It is also worth saying that the reason why what follows is quite technical is because an action is *not* a group homomorphism: $G \times X$ is not even a group.
Now, for the proof proper:

(a) Note that every left action of G on X is, by definition, given by a function $\alpha\colon G \times X \to X$, so that $\mathrm{Act}(G, X) \subseteq \mathrm{Hom}_{\mathbf{Set}}(G \times X, X)$.

---

[1]In fact, currying is precisely Hom-$\otimes$ adjunction (2.7.2) for the closed monoidal category **Set**, where Homs are functions and $\otimes = \times$, the Cartesian product of sets. Later, when we have linearized everything, we will again see Hom-$\otimes$ adjunction playing a similar role.

(b) By the principle of currying, we may define

$$\kappa \colon \mathrm{Act}(G, X) \to \mathrm{Hom}_{\mathsf{Set}}(G, \mathrm{Hom}_{\mathsf{Set}}(X, X))$$

by $\kappa(\alpha)(g) = \alpha(g, -)$.

The "blank" notation, $\alpha(g, -)$, means "think of the blank as a place we can put an element" (in this case x) so that we have a function $\alpha(g, -) \colon X \to X$ given by $\alpha(g, -)(x) = \alpha(g, x)$. So we could (and perhaps should) write $\kappa(\alpha)(g)(x) = \alpha(g, x)$. You need to trace through carefully to see why this is the right thing to do—lots of things here are functions and you just need to keep evaluating them on the right elements! (Arguably, in the same spirit we should have just written $\kappa(\alpha) = \alpha(-, -)$ when defining κ, but more than one blank can become confusing, so we went for the middle ground.)

(c) Unpacking a little further, we see that if we write $\triangleright$ rather than α as before, then $\kappa(\alpha)(g) \in \mathrm{Hom}_{\mathsf{Set}}(X, X)$ sends x to $g \triangleright x$. Notice(!) that

$$\left(\kappa(\alpha)(g^{-1}) \circ \kappa(\alpha)(g)\right)(x) = \kappa(\alpha)(g^{-1})(g \triangleright x) = g^{-1} \triangleright (g \triangleright x) = x$$

and likewise with g^{-1} and g interchanged. So $\kappa(\alpha)(g)$ is actually a bijection, i.e. for all $g \in G$, $\kappa(\alpha)(g) \in \mathrm{Aut}_{\mathsf{Set}}(X)$. This only works because α is a group action; it is not a formal consequence of currying arbitrary functions.

So by restricting the codomain of each $\kappa(\alpha)$, we can define a function

$$\hat{\kappa} \colon \mathrm{Act}(G, X) \to \mathrm{Hom}_{\mathsf{Set}}(G, \mathrm{Aut}_{\mathsf{Set}}(X)), \hat{\kappa}(\alpha) = \kappa(\alpha)|^{\mathrm{Aut}_{\mathsf{Set}}(X)}.$$

(d) We claim that $\hat{\kappa}(\alpha)$ is a group homomorphism. We have $\hat{\kappa}(\alpha)(e)(x) = \alpha(e, x) = x$ for all $x \in X$ so $\hat{\kappa}(\alpha)(e) = \mathrm{id}_X$. Also,

$$\hat{\kappa}(\alpha)(gh)(x) = gh \triangleright x = g \triangleright (h \triangleright x) = (\hat{\kappa}(\alpha)(g) \circ \hat{\kappa}(\alpha)(h))(x)$$

for all $x \in X$, so $\hat{\kappa}(\alpha)$ is a group homomorphism. Note that the axioms of a group action are exactly what is needed for this.

Hence, $\mathrm{Im}\,\hat{\kappa} \subseteq \mathrm{Hom}_{\mathsf{Grp}}(G, \mathrm{Aut}_{\mathsf{Set}}(X)) = \mathrm{Rep}(G, X)$.

(e) It remains to prove that $\hat{\kappa}$ is a bijection, which we will do by showing that it is invertible.

Given a G-representation $\rho \colon G \to \mathrm{Aut}_{\mathsf{Set}}(X)$, define $\alpha_\rho \colon G \times X \to X$ by $\alpha_\rho(g, x) = \rho(g)(x)$. Then for all $x \in X$

$$\alpha_\rho(e, x) = \rho(e)(x) = \mathrm{id}_X(x) = x$$

and

$$\alpha_\rho(g_1 g_2, x) = \rho(g_1 g_2)(x)$$
$$= (\rho(g_1) \circ \rho(g_2))(x)$$
$$= \rho(g_1)(\alpha_\rho(g_2, x))$$
$$= \alpha_\rho(g_1, \alpha_\rho(g_2, x))$$

for all $g_1, g_2 \in G$ and so we see that $\alpha_\rho \in \mathrm{Act}(G, X)$.

Now define

$$\hat{\lambda} \colon \mathrm{Hom}_{\mathbf{Grp}}(G, \mathrm{Aut}_{\mathbf{Set}}(X)) \to \mathrm{Act}(G, X), \ \hat{\lambda}(\rho) = \alpha_\rho.$$

Then

$$\hat{\kappa}(\hat{\lambda}(\rho))(g)(x) = \hat{\kappa}(\alpha_\rho)(g)(x)$$
$$= \alpha_\rho(g, x)$$
$$= \rho(g)(x)$$

and

$$\hat{\lambda}(\hat{\kappa}(\alpha))(g)(x) = \hat{\lambda}(\alpha(g, -))(x)$$
$$= \alpha_{\alpha(g,-)}(x)$$
$$= \alpha(g, x)$$

and we see that $\hat{\kappa}$ and $\hat{\lambda}$ are inverse to each other. $\square$

Challenge 16. Show that if $f \colon \alpha \to \beta$ is a morphism of left actions of G on X, then f induces a morphism of G-representations $\hat{\kappa}(f) \colon \hat{\kappa}(\alpha) \to \hat{\kappa}(\beta)$. Furthermore $\hat{\kappa}$ respects composition: if f, g are morphisms of left actions, $\hat{\kappa}(g \circ f) = \hat{\kappa}(g) \circ \hat{\kappa}(f)$. That is, $\hat{\kappa}$ is a functor from the category $\mathrm{Act}(G, X)$ to the category $\mathrm{Rep}(G, X)$.

By similarly enhancing $\hat{\lambda}$ to a functor in the opposite direction, prove that $\mathrm{Act}(G, X)$ and $\mathrm{Rep}(G, X)$ are isomorphic categories.

Examples 3.1.8. Let us revisit the examples of actions from Examples 3.1.2, expressing them as representations via the proposition.

(a) For $G = \mathrm{GL}_2(\mathbb{R})$ acting on $\mathbb{R}^2$, the representation corresponding to $\left(\begin{smallmatrix} a & b \\ c & d \end{smallmatrix}\right) \triangleright \left(\begin{smallmatrix} x \\ y \end{smallmatrix}\right) = \left(\begin{smallmatrix} ax+by \\ cx+dy \end{smallmatrix}\right)$ is $\rho \colon \mathrm{GL}_2(\mathbb{R}) \to \mathrm{Aut}_{\mathbf{Set}}(\mathbb{R}^2)$ with $\rho(M)$ the function sending $v \in \mathbb{R}^2$ to Mv. But then $\rho(M)$ is the (invertible) linear transformation M represents; this is the *natural* representation of $\mathrm{GL}_2(\mathbb{R})$.

Note (as this is relevant later!) that $\mathrm{Aut}_{\mathbf{Set}}(\mathbb{R}^2)$ *contains* $\mathrm{GL}_2(\mathbb{R})$ but is not equal to it; there are set bijections of $\mathbb{R}^2$ with itself that are not linear, let alone invertible.

(b) For $G = S_n$ acting on $X = \{1, 2, \ldots, n\}$ with $\sigma \triangleright r = \sigma(r)$, we have that $\mathrm{Aut}_{\mathsf{Set}}(X) = S_n$ and the corresponding representation $\rho \colon S_n \to S_n$ is the identity map. This is the *natural* representation of S_n.

Furthermore, any subgroup H of S_n gives rise to a representation $\rho \colon H \hookrightarrow S_n$ in the same way. Since any finite group G is isomorphic to a subgroup of S_G (this is Cayley's theorem, 1.2.72), every finite group has a *permutation* representation $\rho \colon G \hookrightarrow S_G$.

(c) For G acting on itself via $g \triangleright h = gh$, the associated representation is $\rho \colon G \to \mathrm{Aut}_{\mathsf{Set}}(G)$ with $\rho(g)(h) = L_g(h) = gh$, i.e. G is represented by its left multiplication maps. However $\mathrm{Aut}_{\mathsf{Set}}(G) = S_G$ by definition, so in fact this is just the permutation representation!

Note that in general we may have $\mathrm{Aut}_{\mathsf{Grp}}(G) \subsetneq \mathrm{Aut}_{\mathsf{Set}}(G)$. In this particular example, each $\rho(g)$ is actually a group homomorphism and $\mathrm{Im}\,\rho \subseteq \mathrm{Aut}_{\mathsf{Grp}}(G)$.

(d) Exercise 3.8(a).

(e) Exercise 3.8(b).

We will not give code snippets here but just remind you that we discussed how to create matrix rings and groups as well as permutation groups in SageMath in Chapter 1, if you find yourself needing to do specific calculations. The index of SageMath commands on page 215 will probably be helpful.

3.1.3   Linear representations of groups

As the first examples above might suggest, $\mathrm{Aut}_{\mathsf{Set}}(X)$ can be unwieldy to work with. Also, for many group actions, there is additional structure around that is ignored by looking at just the underlying sets and functions on them. By restricting the class of representations we care about, often we are able to say more.

With this in mind, from this point onward we will concentrate on *linear representations*. That is, we will require the set G acts on to be a vector space V over some field $\mathbb{K}$ and ask that representations take values in $\mathrm{GL}(V) \overset{\mathrm{def}}{=} \mathrm{Aut}_{\mathsf{Vect}_{\mathbb{K}}}(V)$.

Definition 3.1.9. Let $\mathbb{K}$ be a field, V a $\mathbb{K}$-vector space and G a group. A $\mathbb{K}$-*linear representation* of G on V is a group homomorphism $\rho \colon G \to \mathrm{GL}(V)$.

Denote the collection of $\mathbb{K}$-linear representations of G on V by $\mathrm{Rep}_{\mathbb{K}}(G, V)$.

Definition 3.1.10. Let $\rho\colon G \to \mathrm{GL}(V)$ be a $\mathbb{K}$-linear representation of G on V. We say ρ is finite-dimensional if V is finite-dimensional and, in that situation, we define the degree of ρ to be $\deg \rho \overset{\text{def}}{=} \dim V$.

Example 3.1.11. The key first example of a linear representation arises from $\mathrm{id}_{\mathrm{GL}_2(\mathbb{R})}\colon \mathrm{GL}_2(\mathbb{R}) \to \mathrm{GL}_2(\mathbb{R})$ with $\mathrm{id}_{\mathrm{GL}_2(\mathbb{R})}(T) = T$ ($T\colon \mathbb{R}^2 \to \mathbb{R}^2$ being a linear transformation, sending v to Tv).

Indeed, if H is any subgroup of $\mathrm{GL}_2(\mathbb{R})$ then $\rho\colon H \hookrightarrow \mathrm{GL}_2(\mathbb{R})$ (given by the inclusion) is a representation of H, which we call the *natural* representation. Clearly neither 2 nor $\mathbb{R}$ are special: any subgroup H of $\mathrm{GL}(V)$ has a natural representation on V, $\rho\colon H \hookrightarrow \mathrm{GL}(V)$.

Other groups do not come to us as groups of linear transformations (or equivalently matrices), though. So linear representation theory is precisely about finding out in what ways the elements of our given group can be *represented* by linear transformations (matrices), in a way that is compatible with the group structure.

Example 3.1.12. For any group G and any vector space V, there is a representation given by $\rho\colon G \to \mathrm{GL}(V)$, $\rho(g) = I$ for all $g \in G$, where $I\colon V \to V$ is the identity linear transformation. This is the *trivial* representation of G on V.

The trivial representation of degree 1 is an important special case: this is defined by $\mathbb{1}_G\colon G \to \mathrm{GL}_1(\mathbb{K}) = \mathbb{K} \setminus \{0\}$, $\mathbb{1}_G(g) = 1_{\mathbb{K}}$ for all $g \in G$.

Example 3.1.13. Let $G = C_2$ be cyclic of order 2, generated by g, with $g^2 = e$. Let $\mathbb{K} = \mathbb{C}$ and $V = \mathbb{C}^2$. Define $\rho\colon G \to \mathrm{GL}_2(\mathbb{C})$ by

$$\rho(e) = \left(\begin{smallmatrix} 1 & 0 \\ 0 & 1 \end{smallmatrix}\right)$$

and

$$\rho(g) = \left(\begin{smallmatrix} -5 & -12 \\ 2 & 5 \end{smallmatrix}\right).$$

Since $\left(\begin{smallmatrix} -5 & -12 \\ 2 & 5 \end{smallmatrix}\right)^2 = \left(\begin{smallmatrix} 1 & 0 \\ 0 & 1 \end{smallmatrix}\right)$, this is a homomorphism.

Observe two things:

(a) the choices of V and $\left(\begin{smallmatrix} -5 & -12 \\ 2 & 5 \end{smallmatrix}\right)$ are not very significant. For any field $\mathbb{K}$ and any $\mathbb{K}$-vector space V, if we can find $T \in \mathrm{GL}(V)$ such that $T^2 = I$, we have a representation $\rho\colon G \to \mathrm{GL}(V)$, $\rho(g^i) = T^i$ for $i = 0, 1$. (In particular, $\left(\begin{smallmatrix} -5 & -12 \\ -2 & 5 \end{smallmatrix}\right)$ being a 2×2 matrix is a red herring: the 2-ness comes from picking $V \cong \mathbb{C}^2$, not from G being cyclic of order 2.)

(b) Implicit was that it was enough to find a matrix (or linear transformation) satisfying the relations in G: this is indeed valid, since if G has a presentation $G = \langle X \rangle / \mathcal{R}$ it is enough to check that $\rho(r) = I$ for any $r \in \mathcal{R}$ to define a homomorphism. Note that $\rho(e) = I$ is always required, by the definition of a group homomorphism.

Problem 17. Let $G = \langle a \mid a^n = 1 \rangle$ be cyclic of order n, let $\zeta = e^{2\pi i/n} \in \mathbb{C}$ and $\theta = 2\pi/n$. Show that

$$\rho \colon G \to \mathrm{GL}_1(\mathbb{C}) = \mathbb{C}^\times, \; \rho(a^j) = \zeta^j$$

and

$$\tau \colon G \to \mathrm{GL}_2(\mathbb{C}), \; \tau(a^j) = \left(\begin{smallmatrix} \cos\theta & -\sin\theta \\ \sin\theta & \cos\theta \end{smallmatrix} \right)^j$$

are $\mathbb{C}$-linear representations of G.

Problem 18. Show that the dihedral group of order 8,

$$D_8 = \langle a, b \mid a^4 = b^2 = e, bab = a^{-1} \rangle$$

has a $\mathbb{Q}$-linear representation $\rho \colon D_8 \to \mathrm{GL}_2(\mathbb{Q})$ with $\rho(a) = \left(\begin{smallmatrix} 0 & -1 \\ 1 & 0 \end{smallmatrix} \right)$ and $\rho(b) = \left(\begin{smallmatrix} 1 & 0 \\ 0 & -1 \end{smallmatrix} \right)$.

Solution. We will use SageMath to help us check this. The code

```
1   GL2=GL(2,QQ);
2   A=GL2([[0,-1],[1,0]]);
3   B=GL2([[1,0],[0,-1]]);
4   I=GL2.one();
5   A^4==I,B^2==I,B*A*B*A==I
```

returns (True, True, True). Notice that to avoid having to compute $\rho(a)^{-1} = A^{-1}$ we re-wrote the final relation as $baba = e$ and checked this.

An important property of a representation is that of being faithful.

Definition 3.1.14. A $\mathbb{K}$-linear representation $\rho \colon G \to \mathrm{GL}(V)$ is *faithful* if $\mathrm{Ker}\,\rho = \{e\}$.

By the First Isomorphism Theorem for Groups (1.2.9), $\mathrm{Ker}\,\rho = \{e\}$ implies $\mathrm{Im}\,\rho \cong G$; that is, a faithful representation puts a "faithful" copy of G into $\mathrm{GL}(V)$. Every group has a faithful representation, by the following construction.

Definition 3.1.15. Let G be a group. Let $\mathbb{K}[G]$ be the $\mathbb{K}$-vector space[2] with basis G. Let $L_g \in \mathrm{GL}(\mathbb{K}[G])$ be the linear map defined on the basis of $\mathbb{K}[G]$ by $L_g(h) = gh$.

The *regular* representation of G is $\rho_{\mathrm{reg}} \colon G \to \mathrm{GL}(\mathbb{K}[G])$, $\rho_{\mathrm{reg}}(g) = L_g$.

---

[2]Recall from Section 2.4 that this notation means the free vector space on G. This notation will be reused and expanded on later, so we use it from the start for consistency.

That this is a representation follows from observing that

$$(L_g \circ L_h)(k) = L_g(hk) = ghk = L_{gh}(k).$$

Furthermore, $\operatorname{Ker} \rho_{\text{reg}} = \{g \in G \mid \rho_{\text{reg}}(g) = I_{\mathbb{K}[G]}\}$ but we can see that $L_g = I_{\mathbb{K}[G]}$ if and only if $g = e$, so ρ_{reg} is faithful.

Example 3.1.16. Let $G = C_4 = \langle a \mid a^4 = e \rangle = \{e, a, a^2, a^3\}$. Then an element in $\mathbb{K}[C_4]$ has the form $\alpha e + \beta a + \gamma a^2 + \delta a^3$ for $\alpha, \beta, \gamma, \delta \in \mathbb{K}$. We know that L_e is the identity, since $L_e(h) = eh = h$ for all $h \in G$.

Now, $L_a(\alpha e + \beta a + \gamma a^2 + \delta a^3) = \alpha a + \beta a^2 + \gamma a^3 + \delta e$ (since $a^4 = e$), from which we see that L_a is represented by the matrix

$$\begin{pmatrix} 0 & 0 & 0 & 1 \\ 1 & 0 & 0 & 0 \\ 0 & 1 & 0 & 0 \\ 0 & 0 & 1 & 0 \end{pmatrix}.$$

By the definition of a representation, $\rho(a^j) = \rho(a)^j$, from which we can deduce the representing matrices for a^2 and a^3.

3.1.4   Equivalence of representations

Let $\varphi \colon V \xrightarrow{\sim} W$ be an isomorphism[3] of $\mathbb{K}$-vector spaces. Since $\operatorname{GL}(V)$ encodes the (linear) symmetries of V, we would expect a close relationship with $\operatorname{GL}(W)$ and indeed, these are isomorphic groups.

Defining

$$f \colon \operatorname{GL}(V) \to \operatorname{GL}(W), \quad f(T) = \varphi \circ T \circ \varphi^{-1}$$

and

$$g \colon \operatorname{GL}(W) \to \operatorname{GL}(V), \quad g(U) = \varphi^{-1} \circ U \circ \varphi,$$

it is straightforward to check that these are inverse homomorphisms.

Then a representation $\rho \colon G \to \operatorname{GL}(V)$ has a counterpart representation $f \circ \rho \colon G \to \operatorname{GL}(W)$ and vice versa, but these are not really "different" representations—we have just changed the underlying vector space for an isomorphic one. So let us say two such representations are equivalent.

Definition 3.1.17. We say that two $\mathbb{K}$-linear representations $\rho \colon G \to \operatorname{GL}(V)$ and $\sigma \colon G \to \operatorname{GL}(W)$ are *equivalent* if there exists a $\mathbb{K}$-vector space isomorphism $\varphi \colon V \to W$ such that

$$
\begin{array}{ccc}
G & \xrightarrow{\ \rho\ } & \operatorname{GL}(V) \\
{\scriptstyle \mathrm{id}_G}\downarrow & & \downarrow{\scriptstyle \hat{\varphi}} \\
G & \xrightarrow[\ \sigma\ ]{} & \operatorname{GL}(W)
\end{array}
$$

commutes, where $\hat{\varphi} \colon \operatorname{GL}(V) \to \operatorname{GL}(W)$ is defined by $\hat{\varphi}(T) = \varphi \circ T \circ \varphi^{-1}$.

---

[3]The added "$\sim$" on the arrow in the specification of a map that is required to be an isomorphism is common but not universal usage. We use it here as part of our attempt to familiarize readers with how representation theorists often write, sometimes without comment.

Since sometimes we prefer more concrete conditions to check, let us unpack this a bit. Choose a basis $\mathcal{B}_V$ for V and a basis $\mathcal{B}_W$ for W.

First, considering $\varphi \colon V \xrightarrow{\sim} W$ as a map from V to W is equivalent to considering the induced map between $(V, \mathcal{B}_V)$ and $(V, \mathcal{B}'_V = \varphi^{-1}(\mathcal{B}_W))$. Then $\hat{\varphi}$ represents the change of basis matrix $[I]^{\mathcal{B}_V}_{\mathcal{B}'_V}$, i.e. the identity map written with respect to $\mathcal{B}_V$ and $\mathcal{B}'_V = \varphi^{-1}(\mathcal{B}_W)$, since a matrix $[T]^{\mathcal{B}}_{\mathcal{B}}$ becomes $[I]^{\mathcal{B}}_{\mathcal{B}'}[T]^{\mathcal{B}}_{\mathcal{B}}[I]^{\mathcal{B}'}_{\mathcal{B}} = [I]^{\mathcal{B}}_{\mathcal{B}'}[T]^{\mathcal{B}}_{\mathcal{B}}([I]^{\mathcal{B}}_{\mathcal{B}'})^{-1}$.

In other words, two group representations are equivalent if for every $g \in G$, $\rho(g)$ and $\sigma(g)$ are similar matrices, with respect to the *same* similarity matrix for every g.

3.1.5 Subrepresentations

We can try to relax the conditions in the definition of equivalence. Let us assume we have an injective linear map $i \colon W \hookrightarrow V$. Then we cannot simply construct $\hat{i}$ as we did $\hat{\varphi}$, since elements of V do not necessarily have a pre-image in W (if i is not surjective). However we can insist that the elements we care about do.

In order to make the following construction, we have to restrict to W actually being a subspace of V (unless we want a mountain of notation and complications). Since an injective map $i \colon W \hookrightarrow V$ has $W \cong \operatorname{Im} i \leq V$ and we have just dealt with isomorphism in the previous section, this is not a major issue.

Definition 3.1.18. Let $\rho \colon G \to \operatorname{GL}(V)$ and $W \leq V$. We say that W is a *G-invariant* subspace if $\rho(G)(W) \leq W$, i.e. for all $g \in G$, $w \in W$ we have $\rho(g)(w) \in W$.

Here, $\rho(G)$ is convenient alternative notation for $\operatorname{Im} \rho$.

Definition 3.1.19. Let $\rho \colon G \to \operatorname{GL}(V)$ be a representation of G and let $W \leq V$ be a G-invariant subspace. Let $i \colon W \hookrightarrow V$ be the associated injective linear map given by $i(w) = w$ for all $w \in W$.

We define $\hat{i} \colon \operatorname{Im} \rho \to \operatorname{GL}(W)$, $\hat{i}(T) = i^{-1} \circ T \circ i$ and $\rho_W \colon G \to \operatorname{GL}(W)$, $\rho_W = \hat{i} \circ \rho$. We say that ρ_W is the *subrepresentation* of ρ associated to W.

Note that the condition of W being G-invariant is taken with respect to the representation $\rho \colon G \to \operatorname{GL}(V)$; it does not only depend on V, in particular. For a given representation, not every subspace of V gives rise to a subrepresentation. Conversely, a subspace W of V might give rise to a subrepresentation with respect to ρ but not with respect to a different representation τ.

In particular, some representations only have the subrepresentations coming from $W = 0$ and $W = V$ (noting that any representation has these subrepresentations; exercise). Such representations are called *irreducible*. If this seems complicated, it is. Part of what we are aiming for is to swap representations for objects where these notions are more natural.

3.1.6   Quotient representations

These should be associated to surjections $V \twoheadrightarrow W$. But rather than work this out in detail for representations, we will wait until we have our better language to do this.

3.1.7   Direct sums

Given V and W, we may form their direct sum, $V \oplus W$. We have a natural map $\delta \colon \mathrm{GL}(V) \times \mathrm{GL}(W) \to \mathrm{GL}(V \oplus W)$, given by $\delta(T, U) = T \oplus U$. The direct sum of linear maps is what you would expect: it sends $(v, w) \in V \oplus W$ to (Tv, Uw); on matrices this is the "block diagonal sum" operation, with $M \oplus N = \left(\begin{smallmatrix} M & 0 \\ 0 & N \end{smallmatrix} \right)$.

Also, let $\Delta \colon G \to G \times G$ be the homomorphism $\Delta(g) = (g, g)$; this is often called the diagonal map.

Definition 3.1.20. Let $\rho \colon G \to \mathrm{GL}(V)$, $\tau \colon G \to \mathrm{GL}(W)$ be representations of G. The *direct sum* $\rho \oplus \tau$ is the representation $\rho \oplus \tau \colon G \to \mathrm{GL}(V \oplus W)$ given by $\rho \oplus \tau = \delta \circ (\rho \times \tau) \circ \Delta$.

That is,

$$(\rho \oplus \tau)(g) = (\delta \circ (\rho \times \tau) \circ \Delta)(g) = \delta \circ (\rho \times \tau)(g, g) = \delta(\rho(g), \tau(g)) = \rho(g) \oplus \tau(g).$$

At this point, we will pause our study of representations of groups. Next we look at the analogous ideas for algebras.

3.2   Representations of algebras

Algebras have representations too but first, we need to work out where a representation should "land". That is, which algebra should replace $\mathrm{GL}(V)$? Well, $\mathrm{GL}(V)$ is used in the group setting because it is the group of invertible linear maps from the vector space to itself. Now that we don't need to get a group, we can relax the invertibility condition and take *all* linear maps from V to itself. It turns out that this is naturally an algebra.

Definition 3.2.1. Let V be a $\mathbb{K}$-vector space. The *endomorphism algebra* of V, $\mathrm{End}_{\mathbb{K}}(V)$, is the $\mathbb{K}$-algebra with vector space $\mathrm{End}_{\mathbb{K}}(V) = \mathrm{Hom}_{\mathbf{Vect}_{\mathbb{K}}}(V, V)$, $m = \circ$ (composition of linear maps) and $u(\lambda) = \lambda I_V$ for all $\lambda \in \mathbb{K}$.

Problem 19. Verify the claim implicit in the definition.

Definition 3.2.2. Let $A = (A, +, \cdot, \times)$ be a $\mathbb{K}$-algebra and V a $\mathbb{K}$-vector space. A ($\mathbb{K}$-linear) *representation* of A on V is defined to be an algebra homomorphism $\rho \colon A \to \mathrm{End}_{\mathbb{K}}(V)$. Denote the collection of representations of A on V by $\mathrm{Rep}_{\mathbb{K}}(A, V)$.

We can define equivalence of representations, subrepresentations and direct sums completely analogously to the group case. Rather than do this explicitly, though, we will turn to representations of quivers next and only after that return to show how the module perspective can drastically simplify matters.

3.3 Representations of quivers

Recall that a quiver $\mathcal{Q}$ is defined by its vertices $\text{Vert}(\mathcal{Q})$ and arrows $\mathcal{Q}(v, w)$ (for $v, w \in \text{Vert}(\mathcal{Q})$). We also defined (Definition 2.1.3) the associated path category, $\mathcal{P}(\mathcal{Q})$, with objects $\text{Vert}(\mathcal{Q})$ and morphisms all finite paths in $\mathcal{Q}$.

The latter enables us to give a slick definition of a representation of a quiver. We will again stick to linear representations.

Definition 3.3.1. Let $\mathcal{Q}$ be a quiver and $\mathbb{K}$ a field. A $\mathbb{K}$-linear *representation* of $\mathcal{Q}$ is a functor $F \colon \mathcal{P}(\mathcal{Q}) \to \mathbb{K}\text{-Mod}$.

Let us unpack this a bit, so that it looks both more comprehensible and also like the definition you are likely to find in other texts. Firstly, recall that $\mathbb{K}\text{-Mod} \equiv \textbf{Vect}_{\mathbb{K}}$ is the category of $\mathbb{K}$-vector spaces. So the first bit of information in a quiver representation is the choice of a vector space Fv for each $v \in \text{Vert}(\mathcal{Q})$.

Next, for each morphism $p \in \mathcal{P}(\mathcal{Q})(v, w)$ in $\mathcal{P}(\mathcal{Q})$, i.e. a path from v to w, F gives a $\mathbb{K}$-linear map $Fp \colon Fv \to Fw$. This is true in particular for paths of length 1, which are just arrows in $\mathcal{Q}$: if $\alpha \in \mathcal{Q}(v, w)$ is an arrow from v to w, we have a linear map $F\alpha \colon Fv \to Fw$. The functoriality means that for

$$p = (v_0, \alpha_0, v_1, \alpha_1, \dots, v_{n-1}, \alpha_{n-1}, v_n)$$

we must have

$$Fp = F\alpha_{n-1} \circ \cdots \circ F\alpha_1 \circ F\alpha_0.$$

In particular, if p, q are paths such that $q \circ p$ is defined, then $F(q \circ p) = Fq \circ Fp$. This tells us that the values of F on paths are determined by its values on arrows.

So a linear representation of a quiver $\mathcal{Q}$ is the choice of a vector space for each vertex and linear maps for each arrow. As we will see later, a useful piece of information about a representation is that of the dimension of the vector space at each vertex; this is very rarely enough to determine the representation uniquely, though.

Definition 3.3.2. Let $\mathcal{Q}$ be a quiver and F a representation of $\mathcal{Q}$. The function $\underline{\dim}\, F \colon \text{Vert}(\mathcal{Q}) \to \mathbb{N}$ defined by $(\underline{\dim}\, F)(v) = \dim Fv$ is called the *dimension function* of F.

When $\mathcal{Q}$ is a finite quiver, the tuple $((\underline{\dim}\, F)(v))_{v \in \text{Vert}(\mathcal{Q})}$ is more often called the *dimension vector* of F.

One advantage of this categorical definition (and we point to some others in Section 6.4) is that it is immediately clear what a morphism of representations should be.

Definition 3.3.3. Let Q be a quiver, $\mathbb{K}$ a field and $F, G \colon \mathcal{P}(Q) \to \mathbb{K}$-Mod linear representations of Q. Then a *morphism* of representations is a natural transformation of functors, $\alpha \colon F \Rightarrow G$.

Again, we should unpack this. A natural transformation is given by a choice of morphism in the target category for each object of the source category such that a certain square commutes. In our case, for $v \in \text{Vert}(Q)$ we have a component $\alpha_v \colon Fv \to Gv$, i.e. a linear map from the vector space at the vertex v given by F to that at the same vertex but given by G. The naturality condition says that for any morphism in the path category, i.e. any path p, from v to w, we must have that $\alpha_w \circ Fp = Gp \circ \alpha_v$. As above, due to the construction of the path category, it is enough to check this on arrows (paths of length 1).

Let us see this in an example.

Example 3.3.4. Let $Q = A_3$ be the quiver $1 \longrightarrow 2 \longrightarrow 3$. The following are representations of this quiver:

$$E \qquad 0 \xrightarrow{\ 0\ } \mathbb{K} \xrightarrow{\ 0\ } 0$$

$$F \qquad 0 \xrightarrow{\ 0\ } \mathbb{K} \xrightarrow{\ \text{id}_\mathbb{K}\ } \mathbb{K}$$

$$G \qquad \mathbb{K} \xrightarrow{\ \text{id}_\mathbb{K}\ } \mathbb{K} \xrightarrow{\ \text{id}_\mathbb{K}\ } \mathbb{K}$$

The dimension vector of E is $(0, 1, 0)$, that of F is $(0, 1, 1)$ and that of G is $(1, 1, 1)$.

There is a morphism of representations $\alpha \colon F \Rightarrow G$ given by

$$
\begin{array}{ccccccc}
F & 0 & \xrightarrow{\ 0\ } & \mathbb{K} & \xrightarrow{\ \text{id}_\mathbb{K}\ } & \mathbb{K} \\
\alpha\Big\Downarrow & \Big\downarrow 0 & & \Big\downarrow \text{id}_\mathbb{K} & & \Big\downarrow \text{id}_\mathbb{K} \\
G & \mathbb{K} & \xrightarrow{\ \text{id}_\mathbb{K}\ } & \mathbb{K} & \xrightarrow{\ \text{id}_\mathbb{K}\ } & \mathbb{K}
\end{array}
$$

One can see immediately in this example the squares here commute.

Let $\lambda \in \mathbb{K}$. There is an endomorphism β_λ of G given by

$$
\begin{array}{ccccccc}
G & \mathbb{K} & \xrightarrow{\ \text{id}_\mathbb{K}\ } & \mathbb{K} & \xrightarrow{\ \text{id}_\mathbb{K}\ } & \mathbb{K} \\
\beta_\lambda\Big\Downarrow & \Big\downarrow \lambda \cdot \text{id}_\mathbb{K} & & \Big\downarrow \lambda \cdot \text{id}_\mathbb{K} & & \Big\downarrow \lambda \cdot \text{id}_\mathbb{K} \\
G & \mathbb{K} & \xrightarrow{\ \text{id}_\mathbb{K}\ } & \mathbb{K} & \xrightarrow{\ \text{id}_\mathbb{K}\ } & \mathbb{K}
\end{array}
$$

For $\lambda = 1_\mathbb{K}$ this is the identity morphism from G to itself.

However

$$
\begin{array}{ccccc}
E & 0 & \xrightarrow{\ 0\ } & \mathbb{K} & \xrightarrow{\ 0\ } & 0 \\
& \downarrow{\scriptstyle 0} & & \downarrow{\scriptstyle \mathrm{id}_\mathbb{K}} & & \downarrow{\scriptstyle 0} \\
G & \mathbb{K} & \xrightarrow{\ \mathrm{id}_\mathbb{K}\ } & \mathbb{K} & \xrightarrow{\ \mathrm{id}_\mathbb{K}\ } & \mathbb{K}
\end{array}
$$

is not a morphism from E to G since the right-hand square does not commute.

We can use SageMath to verify these claims: work through the code below, evaluating as you go to see what the results are. This goes via the construction path_semigroup, which is (give or take) SageMath's way of implementing the path category. In particular, try to think about why the definitions of E, F and G match those above.

```
1  Q = DiGraph({1:{2:['e12']},2:{3:['e23']}})
2  PQ = Q.path_semigroup()
3  E = PQ.representation(QQ,{1: QQ^0, 2: QQ^1, 3: QQ
      ^0},{(1,2,'e12'): [], (2,3,'e23'): []})
4  F = PQ.representation(QQ,{1: QQ^0, 2: QQ^1, 3: QQ
      ^1},{(1,2,'e12'): [], (2,3,'e23'): [1]})
5  G = PQ.representation(QQ,{1: QQ^1, 2: QQ^1, 3: QQ
      ^1},{(1,2,'e12'): [1], (2,3,'e23'): [1]})
6  F.dimension()
7  G.dimension_vector()
8  alpha = F.hom({1:[], 2:[1], 3:[1]}, G)
9  beta3 = G.hom({1:[3], 2:[3], 3:[3]}, G)
10 notahom = E.hom({1:[], 2:[1], 3:[]}, G)
```

Note that we use $\mathbb{K} = \mathbb{Q}$, for concreteness.

Since we have some objects and some morphisms between them, we could have a category.

Definition 3.3.5. Let $\mathcal{Q}$ be a quiver and $\mathbb{K}$ a field. Let $\mathrm{Rep}(\mathcal{Q})$ denote the category with objects $\mathbb{K}$-linear representations of $\mathcal{Q}$ and morphisms being morphisms of representations.

You might object that it is not clear that this is a category, and might not fancy checking all the conditions. Fear not: all will be revealed[4] in Section 6.4.

We now "know" how to define many of the representation-theoretic concepts for quivers, by following the same recipes but in the category $\mathrm{Rep}(\mathcal{Q})$.

---

[4]Spoiler: because we set things up in terms of categorical definitions, we get this essentially for free.

For example, equivalence of representations is just isomorphism in this category, i.e. $\alpha\colon F \Rightarrow G$ a natural isomorphism (recall that this means that every component α_v is an isomorphism of vector spaces, so an invertible linear map, or equivalently an element of $\mathrm{GL}(Fv)$).

So β_λ is an automorphism of G (an isomorphism of G with itself) in the above example provided $\lambda \neq 0$. Since $F1 = 0 \not\cong \mathbb{K} = G1$, F cannot be isomorphic to G.

Using categorical definitions, we could define subrepresentations (exercise: make sense of the claim that F is a subrepresentation of G), quotients and so on. Again, we will not do this now, but come back to these when we are ready to discuss the counterpart to representations, namely modules for quivers.

3.E   Exercises

Exercise 3.1 (Problem 10). Write down the definition of a right action.

Exercise 3.2 (Problem 11). Write down and prove (using the axioms) a mathematical statement that formalizes the statement "we want a symmetry to be reversible".

Exercise 3.3 (Problem 12). Describe the orbits and stabilizers for the following actions:

(a) $\mathrm{GL}_2(\mathbb{R})$ acting on $\mathbb{R}^2$, $\left(\begin{smallmatrix} a & b \\ c & d \end{smallmatrix}\right) \triangleright \left(\begin{smallmatrix} x \\ y \end{smallmatrix}\right) = \left(\begin{smallmatrix} ax+by \\ cx+dy \end{smallmatrix}\right)$;

(b) S_n acting on $\{1, 2, \ldots, n\}$, $\sigma \triangleright r = \sigma(r)$;

(c) G acting on itself via $g \triangleright h = gh$;

(d) G acting on itself via $g \triangleright h = ghg^{-1}$;

(e) G acting on the left cosets of a subgroup H via $g \triangleright kH = gkH$.

Exercise 3.4 (Problem 13). Let $\triangleright\colon G \times X \to X$ be a left action of G on X. Prove that the orbits partition X and that the stabilizer of a point $x \in X$ is a subgroup of G.

Exercise 3.5 (Challenge 14). How would you define a *morphism* between two G-actions? That is, given $\alpha\colon G \times X \to X$ and $\beta\colon G \times Y \to Y$, what would be an appropriate notion of a morphism $f\colon \alpha \to \beta$?

Can you show that the collection $\mathrm{Act}(G, X)$ of (left) actions of G on X with your notion of morphism is a category?

Exercise 3.6 (Challenge 15). Show that $\mathrm{Rep}(G, X)$ with objects G-representations and morphisms of G-representations is a category.

Exercise 3.7 (Challenge 16; hard!). Show that if $f: \alpha \to \beta$ is a morphism of left actions of G on X, then f induces a morphism of G-representations $\hat{\kappa}(f): \hat{\kappa}(\alpha) \to \hat{\kappa}(\beta)$. Furthermore $\hat{\kappa}$ respects composition: if f, g are morphisms of left actions, $\hat{\kappa}(g \circ f) = \hat{\kappa}(g) \circ \hat{\kappa}(f)$. That is, $\hat{\kappa}$ is a functor from the category $\mathrm{Act}(G, X)$ to the category $\mathrm{Rep}(G, X)$.

By similarly enhancing $\hat{\lambda}$ to a functor in the opposite direction, prove that $\mathrm{Act}(G, X)$ and $\mathrm{Rep}(G, X)$ are isomorphic categories.

Exercise 3.8. For a group G, express the following actions as representations:

(a) G acting on itself via $g \triangleright h = ghg^{-1}$;

(b) G acting on the left cosets of a subgroup H via $g \triangleright kH = gkH$.

Exercise 3.9. Let $G = \langle a \mid a^2 = 1 \rangle$ be a group of order 2. Find $\lambda \in \mathbb{C}$ such that the function $\rho: G \to \mathrm{GL}_2(\mathbb{C})$,

$$\rho(a^j) = \begin{pmatrix} 10 & -11 \\ \lambda & -10 \end{pmatrix}^j,$$

for $j = 0, 1$, is a representation of G.

Exercise 3.10 (Problem 17). Let $G = \langle a \mid a^n = 1 \rangle$ be cyclic of order n, let $\zeta = e^{2\pi i/n} \in \mathbb{C}$ and $\theta = 2\pi/n$. Show that

$$\rho: G \to \mathrm{GL}_1(\mathbb{C}) = \mathbb{C}^\times, \ \rho(a^j) = \zeta^j$$

and

$$\tau: G \to \mathrm{GL}_2(\mathbb{C}), \ \tau(a^j) = \begin{pmatrix} \cos\theta & -\sin\theta \\ \sin\theta & \cos\theta \end{pmatrix}^j$$

are $\mathbb{C}$-linear representations of G.

Exercise 3.11. Let $G = \mathrm{GL}_n(\mathbb{K})$ and $H = M_n(\mathbb{K})$. Decide which of the following assertions hold and briefly prove your claims. Recall that $\mathrm{GL}_1(\mathbb{K}) \cong \mathbb{K}^\times$ as multiplicative groups.

(a) $\det: G \to \mathbb{K}^\times$ is a representation of G.

(b) $\det: H \to \mathbb{K}^\times$ is a representation of H.

(c) $\mathrm{tr}: G \to \mathbb{K}^\times$ is a representation of G.

(d) $\mathrm{tr}: H \to \mathbb{K}^\times$ is a representation of H.

(e) $\det: G \to \mathbb{K}^\times$ is a faithful representation of G.

Exercise 3.12. Let $G = D_{12} = \langle a, b \mid a^6 = b^2 = e, bab = a^{-1} \rangle$, and set

$$C = \begin{pmatrix} e^{i\frac{\pi}{3}} & 0 \\ 0 & e^{-i\frac{\pi}{3}} \end{pmatrix}, \qquad D = \begin{pmatrix} 0 & 1 \\ 1 & 0 \end{pmatrix},$$

$$E = \begin{pmatrix} \frac{1}{2} & \frac{\sqrt{3}}{2} \\ -\frac{\sqrt{3}}{2} & \frac{1}{2} \end{pmatrix}, \qquad F = \begin{pmatrix} 1 & 0 \\ 0 & -1 \end{pmatrix}.$$

Show that ρ_1, ρ_2, ρ_3 and ρ_4 are representations of G, where

$$\rho_1(b^i a^j) = D^i C^j,$$
$$\rho_2(b^i a^j) = (-D)^i C^{3j},$$
$$\rho_3(b^i a^j) = D^i(-C)^j,$$
$$\rho_4(b^i a^j) = F^i E^j.$$

Which of these four representations are faithful? Which of them are equivalent to each other?

[*You might find it helpful to have computer assistance for this question. If you do, keep a record of the code you use for future reference.*]

In particular, you may find it helpful to know that you can construct $e^{i\pi/3}$ in SageMath as the element w via the following:

```
1   QQ.<w>  =  CyclotomicField(6)
2   CC(w)
3   w^6  ==  1
```

Exercise 3.13 (Problem 18). Show that the dihedral group of order 8,

$$D_8 = \langle a, b \mid a^4 = b^2 = e, bab = a^{-1} \rangle$$

has a $\mathbb{Q}$-linear representation $\rho: D_8 \to \mathrm{GL}_2(\mathbb{Q})$ with $\rho(a) = \left(\begin{smallmatrix} 0 & -1 \\ 1 & 0 \end{smallmatrix}\right)$ and $\rho(b) = \left(\begin{smallmatrix} 1 & 0 \\ 0 & -1 \end{smallmatrix}\right)$.

Exercise 3.14. Let $n \geq 3$ and $G = \langle a, b \mid a^n = b^2 = e, bab = a^{-1} \rangle$ a dihedral group of order $2n$. Let $\theta = \frac{2\pi}{n}$ and consider the matrices $A = \begin{pmatrix} \cos\theta & -\sin\theta \\ \sin\theta & \cos\theta \end{pmatrix}$ and $B = \begin{pmatrix} 1 & 0 \\ 0 & -1 \end{pmatrix}$.

(a) Show that the function $\rho: G \to \mathrm{GL}_2(\mathbb{C})$, $\rho(a^k b^l) = A^k B^l$ for all k, l is a representation of G.

(b) Prove that ρ is faithful.

(c) Find a basis $\mathcal{B}$ of $\mathbb{C}^2$ such that the matrix $[\rho(a)]_{\mathcal{B}}$ is diagonal.

Exercise 3.15. Two of the following three representations of the cyclic group $C_{12} = \langle a \mid a^{12} = e \rangle$ are equivalent. Find which ones, and justify your answer.

(a) $\rho_1(a) = \begin{pmatrix} 0 & -1 \\ 1 & 0 \end{pmatrix}$

(b) $\rho_2(a) = \begin{pmatrix} 0 & -\omega^2 \\ \omega & 0 \end{pmatrix}$ where $\omega = e^{\frac{2\pi i}{3}}$ is a complex primitive cube root of 1.

(c) $\rho_3(a) = \begin{pmatrix} 0 & -\omega \\ 1 & -\omega^2 \end{pmatrix}$ where $\omega = e^{\frac{2\pi i}{3}}$ is a complex primitive cube root of 1.

Exercise 3.16 (Problem 19). Verify the claim in Definition 3.2.1 that $\text{End}_{\mathbb{K}}(V)$ is an algebra.

Exercise 3.17. Using Example 3.3.4 as a template, write down three morphisms of representations of the quiver A_3.

Chapter 4

Modules

In the last chapter, we first looked at representations of groups on sets, realizing that the minimal assumptions make group actions hard to study collectively. We then linearized, which opens up linear algebra for use in studying linear representations. But again, we were dissatisfied: the constructions one would like to do with an algebraic object are awkward. Subrepresentations are doable but tricky, and quotient representations are just messy. These problems were not resolved by looking at representations of algebras rather than groups: the issues are to do with the nature of representations.

When we considered quiver representations, something else happened. Now we ended up with a rather different flavour of representation and it was not clear how it related to the cases of groups or algebras. That would mean having to develop two separate lots of theory.

We claim that the answer to all these problems is to study modules instead. We will be able to make all the constructions we wanted, and more, and we will be able to leverage the same theory in many different settings.

It may take a little time to become comfortable with the definition of a module but hopefully you will come to realize that it is as natural as the other algebraic structures you have met before. However, in order to maximize the gains from the "do it once and for all" approach, we will need to be slightly more general again.

Specifically, in this chapter, we will give definitions and explore results for fundamental concepts associated to modules over R-algebras, where R is now a commutative unital ring that is not necessarily a field (e.g. $R = \mathbb{Z}$, rather than $R = \mathbb{R}$). We work in this level of generality so that in the following chapter, we can specialize both to $A = R$ and talk about modules over commutative rings (for certain special classes of commutative ring) and also specialize to modules over $\mathbb{K}$-algebras (for certain special classes of $\mathbb{K}$-algebras). This might feel less comfortable than working over a field, but as we will see, we will rarely want to divide by scalars—just multiply by them—so not having inverses will not be a big deal.

4.1   Modules for algebras

Over the next few pages, we will both give the definition of a module and also re-define algebras in the more general setting discussed above. We will emphasize the key features of these objects: in the case of modules, it is the action, and in the case of algebras, it is the multiplication. In particular, *everything will be assumed to be R-linear*, where R is our base commutative unital ring. This means that all our objects will be (at a minimum) Abelian groups with an addition operation, a zero element and subtraction, in the usual way. Although we will have to say at various points that linearity is being preserved, we will place much more emphasis on the new and distinctive features (actions, multiplication) than on the underlying linearity.

For this reason, we will go right back to the beginning and reformulate the definition of a ring in a way that will enable us to give our definitions of modules and algebras over a ring in a coherent way. There are several other reasons we have for doing this. Firstly, axioms for algebraic structures are often given in terms of formulæ that can be awkward to parse and to remember. The approach we will introduce shortly has a more pictorial and dynamic flavour, which you will hopefully find more enlightening. Secondly, it will make it easier for us to apply some of the category theory results from Chapter 2. Thirdly, it ultimately leads to new mathematical ideas, as we will see in Section 6.3 when Hopf algebras are introduced.

4.1.1   A categorical approach to rings

In order to make sense of what we will do next, we recommend looking over what we said about tensor products in Sections 1.4.1 and 2.7, noting in particular Definition 2.7.3 defining $\otimes$, the tensor product of Abelian groups.

Definition 4.1.1. A ring is an algebraic structure (R, m) such that R is an Abelian group and $m \colon R \otimes R \to R$ (a homomorphism of Abelian groups) such that the following diagram commutes:

$$
\begin{array}{ccc}
R \otimes R \otimes R & \xrightarrow{\ m \otimes \mathrm{id}\ } & R \otimes R \\
{\scriptstyle \mathrm{id} \otimes m} \downarrow & & \downarrow {\scriptstyle m} \\
R \otimes R & \xrightarrow[\ \ m\ \ ]{} & R
\end{array}
$$

Here, $(m \otimes \mathrm{id})(r_1 \otimes r_2 \otimes r_3) = m(r_1 \otimes r_2) \otimes r_3 \in R \otimes R$ (extended linearly to sums of such tensors), i.e. we apply m to the first two tensor factors and apply the identity map to the third (or equivalently, leave the third alone).

Notice, though, that our definition avoids explicitly quantifying over elements; this is handled by saying that the maps are equal, since the maps are equal if and only if they take the same values on all possible inputs. This is particularly helpful when we have tensor products around, since then we can avoid having to do as in the previous paragraph and repeatedly say "defined by [*definition*] on elementary tensors and extended linearly".

Notice too that we have deliberately suppressed the addition $+$ on the underlying Abelian group R from the notation. You should be aware that it is there although we don't want to focus on this, but instead highlight the key part, i.e. m.

Now let us try to explain why the above is actually equivalent to our old definition, Definition 1.1.3. Firstly, we start both definitions with R being an Abelian group, $(R, +)$, if we want to be explicit about $+$.

Next, let us compute the two compositions of the maps given in the diagram on a chosen elementary tensor $a \otimes b \otimes c$ of $R \otimes R \otimes R$ (where $a, b, c \in R$). Temporarily, we will write an explicit $\times$ for multiplication, that is, the operation $\times$ is defined by $m(a \otimes b) = a \times b$. Then

- (across then down) $(m \circ (m \otimes \mathrm{id}))(a \otimes b \otimes c) = m((a \times b) \otimes c) = (a \times b) \times c$

- (down then across) $(m \circ (\mathrm{id} \otimes m))(a \otimes b \otimes c) = m(a \otimes (b \times c)) = a \times (b \times c)$

So, the equality of these two maps means that $(a \times b) \times c = a \times (b \times c)$ for all $a, b, c \in A$, that is, the product is associative. (From now on, we will revert to using concatenation to denote multiplication, so will write $m(a \otimes b) = ab$.)

Then as in Remark 1.3.1, the distributive laws are equivalent to left and right multiplication by a fixed element being homomorphisms of the underlying Abelian group. But this is exactly what is encoded in m being a homomorphism from the tensor product: this takes a little work that we will not do explicitly but perhaps you will be able to see a hint about why it should be true in the relations that are imposed in the standard model of a tensor product we constructed on page 59.

As before, this is a definition of a "(not necessarily unital) ring" but we can give a definition of a unital ring in the same spirit as above, as follows.

Definition 4.1.2. Let $R = (R, m)$ be a ring. Then a homomorphism of Abelian group $u \colon \mathbb{Z} \to R$ is a multiplicative identity for R if

commutes. We say that (R, m, u) is a unital ring if $R = (R, m)$ is a ring and u is a multiplicative identity for R.

Here, the maps $\triangleright$ and $\triangleleft$ are the canonical[1] homomorphisms defined by

$$\triangleright(n \otimes r) = nr = \overbrace{r + r + \cdots + r}^{n \text{ times}} = rn = \triangleleft(r \otimes n)$$

Let us unpack this a bit too, to see why it means the same as our earlier definitions. Firstly, any homomorphism of Abelian groups $u\colon \mathbb{Z} \to R$ is completely determined by the value $u(1)$. This is because $\mathbb{Z}$ is cyclic and generated by 1, under addition:

$$n = n1 = \overbrace{1 + 1 + \cdots + 1}^{n \text{ times}}$$

So, we can concentrate on following 1 through our diagram.

Next, from the left-hand triangle,

$$(m \circ (u \otimes \mathrm{id}))(1 \otimes r) = m(u(1) \otimes r) = u(1)r$$

is equal to $\triangleright(1 \otimes r) = 1r = r$. That is, $u(1)r = r$ for all $r \in R$. Similarly, $ru(1) = r$ for all r. So, $u(1)$ has exactly the property to be a multiplicative identity for R; we could define $1_R = u(1)$, if we wanted to simplify notation slightly.

The more categorical Definition 4.1.2 helps us realize that multiplicative identities are chosen, and not necessarily canonical (as is also true of multiplication, in fact). That said, in many examples, there is a natural—nay, canonical—choice of u.

The statement of Lemma 1.4.5 and its proof generalize to this context too. That is, there is an isomorphism $\tau_{VW}\colon V \otimes W \to W \otimes V$ whose value on an elementary tensor $v \otimes w$ is $w \otimes v$, that is, τ_{VW} "flips" the tensor product order.

This applies in particular to a ring R: let $\tau\colon R \otimes R \to R \otimes R$ be the isomorphism induced by $\tau(r_1 \otimes r_2) = r_2 \otimes r_1$ on elementary tensors, extended linearly. Then we may make the following definition.

Definition 4.1.3. Let (R, m) be a ring. We say that R is commutative if $m \circ \tau = m$.

On elementary tensors, this says

$$r_2 r_1 = m(r_2 \otimes r_1) = (m \circ \tau)(r_1 \otimes r_2) = m(r_1 \otimes r_2) = r_1 r_2$$

which is the usual equality expressing commutativity of multiplication. So now we have the definition of a commutative unital ring in this more categorical phrasing.

---

[1]We say a construction is *canonical* to indicate that it arises in a particularly natural way; there is not really a formal definition of what it means to be canonical, which is why some authors dislike it. We use it, as most people do, suggestively.

Lastly, let us also re-express ring homomorphisms in this language, not because we will make especially heavy use of them in this form, but to provide a point of comparison for the new definitions to come soon.

To save space, we will give the definition for unital rings; the definition for not necessarily unital rings is obtained by simply deleting the word "unital" along with references to the maps u_R and u_S and the corresponding diagram.

Definition 4.1.4. Let (R, m_R, u_R) and (S, m_S, u_S) be unital rings. A unital homomorphism of unital rings is a homomorphism of Abelian groups $f \colon R \to S$ such that the following diagrams commute:

(a) (multiplication preserved)

$$
\begin{array}{ccc}
R \otimes R & \xrightarrow{\ f \otimes f\ } & S \otimes S \\
\ \downarrow{\scriptstyle m_R} & & \ \downarrow{\scriptstyle m_S} \\
R & \xrightarrow{\quad f \quad} & S
\end{array}
$$

(b) (multiplicative identity preserved)

$$
\begin{array}{ccc}
R & \xrightarrow{\quad f \quad} & S \\
& {\scriptstyle u_R}\nwarrow \quad \nearrow{\scriptstyle u_S} & \\
& \mathbb{Z} &
\end{array}
$$

Unpacking the first of these like we did before, we have

$$(m_S \circ (f \otimes f))(r_1 \otimes r_2) = m_S(f(r_1) \otimes f(r_2))$$
$$= f(r_1)f(r_2)$$

and

$$(f \circ m_R)(r_1 \otimes r_2) = f(r_1 r_2)$$

so that the first condition means $f(r_1 r_2) = f(r_1)f(r_2)$ for all $r_1, r_2 \in R$.

Similarly, preservation of the multiplicative identity is equivalent to $f(u_R(1)) = u_S(1)$ (i.e. $f(1_R) = 1_S$).

4.1.2   Modules over commutative unital rings

At last, we are ready for our most important definition, that of a module over a commutative unital ring. From this point, essentially everything we do will be to study such modules for different rings and for algebras over rings. For—we claim—this is what representation theory is.

Definition 4.1.5. Let $R = (R, m, u)$ be a commutative unital ring and let M be an Abelian group. We say that the pair $(M, \triangleright)$ is a *left R-module* if $\triangleright \colon R \otimes M \to M$ is a homomorphism of Abelian groups such that the diagrams

$$
\begin{array}{ccc}
R \otimes R \otimes M & \xrightarrow{m \otimes \mathrm{id}} & R \otimes M \\
{\scriptstyle \mathrm{id} \otimes \triangleright} \downarrow & & \downarrow {\scriptstyle \triangleright} \\
R \otimes M & \xrightarrow{\triangleright} & M
\end{array}
$$

and

$$
\begin{array}{ccc}
\mathbb{Z} \otimes M & \xrightarrow{u \otimes \mathrm{id}} & R \otimes M \\
& \searrow \quad \swarrow {\scriptstyle \triangleright} & \\
& M &
\end{array}
$$

commute.

Here, the map $\mathbb{Z} \otimes M \to M$ is that which we used before,[2] namely $n \otimes m \mapsto nm$.

The conditions in this definition can be seen, via similar considerations as above, to be equivalent to:

(a) $(r + s) \triangleright m = (r \triangleright m) + (s \triangleright m)$ for all $r, s \in R, m \in M$;

(b) $r \triangleright (m + n) = (r \triangleright m) + (r \triangleright n)$ for all $r \in R, m, n \in M$;

(c) $r \triangleright (s \triangleright m) = (r \times s) \triangleright m$ for all $r, s \in R, m \in M$; and

(d) $1_R \triangleright m = m$ for all $m \in M$ (where $1_R = u(1)$ as above).

Furthermore, we can define homomorphisms.

Definition 4.1.6. Let $R = (R, m, u)$ be a commutative unital ring and $(M, \triangleright_M)$, $(N, \triangleright_N)$ left R-modules. We say that a map $f \colon M \to N$ is an *R-module homomorphism* if f is a homomorphism of Abelian groups and the following diagram commutes:

$$
\begin{array}{ccc}
R \otimes M & \xrightarrow{\mathrm{id} \otimes f} & R \otimes N \\
{\scriptstyle \triangleright_M} \downarrow & & \downarrow {\scriptstyle \triangleright_N} \\
M & \xrightarrow{f} & N
\end{array}
$$

That is, if $f(r \triangleright_M m) = r \triangleright_N f(m)$ for all $r \in R$ and $m \in M$.

Then there is a category R-Mod, the category of R-modules, whose objects are R-modules and morphisms R-module homomorphisms. For $R = \mathbb{Z}$, we have $\mathbb{Z}$-Mod = **Ab**, the category of Abelian groups.[3]

---

[2]Previously we used $\triangleright$ for this map, but we want to use that symbol for our R-action now, not the $\mathbb{Z}$-action.

[3]It may be best to take this as a valid assertion being made to begin with. When you feel more comfortable with the definitions, think about why it is true.

Some examples are long overdue.

Example 4.1.7. The *zero module* is the module $(\{0\}, \triangleright = 0)$ with $r \triangleright 0 = 0$ for all $r \in R$. We will usually save ourselves some writing by calling this module 0 rather than $\{0\}$. Any other (sub)module not isomorphic to this is called *non-zero*.

Example 4.1.8. Every ring is a $\mathbb{Z}$-module: we said above that $\mathbb{Z}$-modules are precisely Abelian groups and, by definition, every ring is an Abelian group. The left $\mathbb{Z}$-action on R is exactly

$$\triangleright(n \otimes r) = nr = \overbrace{r + r + \cdots + r}^{n \text{ times}}$$

Indeed, as a quick look ahead to a philosophy we will say more about soon, we can think that rings are exactly "the algebraic structures having an associative multiplication *in the category* **Ab**".

Example 4.1.9. Let R be a commutative unital ring. The polynomial R-algebra $R[x]$ in one variable has a module known as the *trivial module* (not to be confused with the zero module!). This is defined to be the module (R, η) with $\eta \colon R[x] \otimes R \to R$, $\eta(1 \otimes r) = r$ and $\eta(x \otimes r) = 0$ for all $r \in R$.

Indeed, this module structure arises from the homomorphism of R-algebras $c \colon R[x] \to R$, $c(p(x)) = p(1)$, sending a polynomial to its constant term.

Problem 20. Check that the above definition of the trivial module is indeed an $R[x]$-module structure.

Remark 4.1.10. When $R = \mathbb{K}$ is a field, we can talk about the dimension of any $\mathbb{K}$-module (i.e. vector space) in the usual way. For R commutative but not a field, this is not as straightforward, and for R non-commutative, things are significantly more complicated. If interested, we recommend searching for "torsion" and "Krull dimension" (see e.g. [GW]).

Example 4.1.11. Let $\mathbb{K}$ be a field, which is in particular a commutative unital ring. Then, looking carefully at the lists of axioms, we see that M is a $\mathbb{K}$-module if and only if M is a $\mathbb{K}$-vector space. Moreover, $\mathbb{K}$-module homomorphisms are exactly $\mathbb{K}$-linear maps.

That is, the category of $\mathbb{K}$-modules, $\mathbb{K}$-Mod, is the category of $\mathbb{K}$-vector spaces, **Vect**$_{\mathbb{K}}$.

Extending our comment above, an algebra over a field is "an algebraic structure having an associative multiplication *in the category* $\mathbb{K}$-Mod = **Vect**$_{\mathbb{K}}$".

Example 4.1.12. Another important example of a module that we always have, for any R, is $_RR = (R, m)$, the *regular* (left) R-module.

That is, we take $M = R$ and $\triangleright = m$ and observe that substituting these into the above diagrams exactly gives us the diagrams expressing associativity and unitarity in our alternative definition of a ring on page 110. Since those diagrams commute by definition of R being a ring, we have that (R, m) is a module.

Since we are only asking for a left module structure, we only need half of the unitarity diagram.

Problem 21. Give a definition in the above style of a right module for a commutative unital ring R and of the right regular module R_R.

4.1.3   Algebras over commutative unital rings

The next step we take is to formalize our loose remarks above about algebraic structures having an associative multiplication on top of some other structure. The examples were the "other structure" being $\mathbb{Z}$-modules and $\mathbb{K}$-vector spaces (i.e. $\mathbb{K}$-modules), so it is natural to expect that this might make sense for modules over a commutative unital ring, since $\mathbb{Z}$ and $\mathbb{K}$ are examples of this. This is indeed what we now do.

To make this definition, we need the tensor product $\otimes_R$ of R-modules, for R a commutative unital ring, whose definition is given in the same way as in Definitions 1.4.3 or 2.7.3 but with $\mathbb{K}$-module (i.e. vector space) or $\mathbb{Z}$-module (i.e. Abelian group) replaced by R-module.

Definition 4.1.13. Let $R = (R, m_R, u_R)$ be a commutative unital ring. An associative *algebra over* R is an algebraic structure (A, m) such that A is an R-module and $m\colon A \otimes_R A \to A$ (a homomorphism of R-modules) such that the following diagram commutes:

$$
\begin{array}{ccc}
A \otimes_R A \otimes_R A & \xrightarrow{m \otimes \mathrm{id}} & A \otimes_R A \\
{\scriptstyle \mathrm{id} \otimes m} \downarrow & & \downarrow {\scriptstyle m} \\
A \otimes_R A & \xrightarrow[\quad m \quad]{} & A
\end{array}
$$

This is a definition of a "(not necessarily unital) algebra", which we now extend to the unital case.

Definition 4.1.14. Let $R = (R, m_R, u_R)$ be a commutative unital ring and $A = (A, m)$ an algebra over R. Then a homomorphism of R-modules $u\colon R \to A$ is a multiplicative identity for A if

$$
\begin{array}{ccccc}
& & A \otimes_R A & & \\
& {\scriptstyle u \otimes \mathrm{id}} \nearrow & \big| & \nwarrow {\scriptstyle \mathrm{id} \otimes u} & \\
R \otimes_R A & & {\scriptstyle m} \big| & & A \otimes_R R \\
& \searrow & \downarrow & \swarrow & \\
& & A & &
\end{array}
$$

commutes. We say that (A, m, u) is a unital associative algebra over R, or *R-algebra* for short, if $A = (A, m)$ is a ring and u is a multiplicative identity for A.

Here, the maps $\triangleright$ and $\triangleleft$ are the canonical homomorphisms defined by the R-module structure: $\triangleright(r \otimes_R a) = r \triangleright a$ and $\triangleleft(a \otimes_R r) = r \triangleright a$. The latter may seem a bit odd, but it does work, because R is commutative. The special case to have in mind is that in a vector space, $v\lambda = \lambda v$ for $\lambda \in \mathbb{K}$ and $v \in V$.

Since R is unital with multiplicative identity $1_R = u_R(1)$ (the image of the integer 1 under u_R), we see that the left-hand triangle tells us that $(m \circ (u \otimes \mathrm{id}))(1_R \otimes r) = m(u(1_R) \otimes r) = u(1_R)r$ is equal to $\triangleright(1_R \otimes r) = 1_R r = r$. That is, $u(1_R)r = r$ for all $r \in R$. Similarly, $ru(1_R) = r$ for all r. So, $u(1_R)$ has exactly the property to be a multiplicative identity for R and indeed we should define $1_A = u(1_R) = (u \circ u_R)(1)$ as notation for the multiplicative identity of A.

As before, there is an R-module isomorphism $\tau_{VW} \colon V \otimes W \to W \otimes V$ whose value on an elementary tensor $v \otimes w$ is $w \otimes v$, that is, τ_{VW} "flips" the tensor product order, where now V and W are R-modules.

This applies in particular to the algebra A: let $\tau \colon A \otimes_R A \to A \otimes_R A$ be the isomorphism induced by $\tau(a_1 \otimes a_2) = a_2 \otimes a_1$ on elementary tensors, extended linearly. Then we may make the following definition.

Definition 4.1.15. Let (A, m) be an associative algebra over a commutative unital ring R. We say that A is commutative if $m \circ \tau = m$.

In terms of examples, there are the usual trivial ones: R is an R-algebra, with $\triangleright = m_R = m$. Also, every ring is a $\mathbb{Z}$-algebra.

This actually bears a little more examination, as follows. If A is an algebra (or even just a ring), its *centre*, $Z(A)$, is the set of elements of A that commute multiplicatively with every element of A:

$$Z(A) = \{z \in A \mid za = az \; \forall a \in A\}.$$

Then for R commutative and unital, an equivalent formulation of the definition of an R-algebra A is to specify a pair (A, φ) where A is a unital ring and $\varphi \colon R \to Z(A)$ is a unital ring homomorphism.

In the examples above, the identity homomorphism $\mathrm{id} \colon R \to Z(R) = R$ shows that the ring R is an R-algebra and the homomorphism induced by $1 \mapsto 1_R$ shows that R is a $\mathbb{Z}$-algebra.

Of course, we would like some non-trivial examples. A good starting point is $R[x]$, the polynomial R-algebra in one variable, or its generalizations to several variables or indeed formal power series.

Other non-trivial examples will come in the following chapter, most notably the group algebra (see 5.2.1) and the path algebra (see 5.3.1).

Remark 4.1.16. Although we want the generality of algebras over rings for Section 5.1, after that we will mainly want algebras over fields. So we re-emphasize here that if $R = \mathbb{K}$ is a field, we recover our earlier definition (Definition 1.5.1) of an algebra over a field and that an algebra over a field is "a vector space with an associative multiplication" and "a ring with an extra scalar multiplication by elements of a field".

By now, hopefully the following will be visibly the natural definition to make of a homomorphism of (associative unital) algebras (over a ring).

Definition 4.1.17. Let $R = (R, m)$ be a commutative unital ring and let (A, m_A, u_A) and (B, m_B, u_B) be unital associative algebras over R. A unital homomorphism of unital associative algebras is a homomorphism of R-modules $f : R \to S$ such that the following diagrams commute:

(a) (multiplication preserved)

$$
\begin{array}{ccc}
A \otimes_R A & \xrightarrow{\ f \otimes f\ } & B \otimes_R B \\
{\scriptstyle m_A}\big\downarrow & & \big\downarrow{\scriptstyle m_B} \\
A & \xrightarrow{\quad f \quad} & B
\end{array}
$$

(b) (multiplicative identity preserved)

$$
\begin{array}{ccc}
A & \xrightarrow{\ f\ } & B \\
& \nwarrow{\scriptstyle u_A} \quad \nearrow{\scriptstyle u_B} & \\
& \mathbb{Z} &
\end{array}
$$

These diagrams precisely say that $f(a_1 a_2) = f(a_1)f(a_2)$ and $f(1_A) = 1_B$, as you might expect. The preservation of addition is encoded in the statement that f is an R-module homomorphism.

For the "not necessarily unital" version, just delete the no-longer relevant parts.

Then we have a category $\mathbf{Alg}_R$ of associative algebras over R; as with rings, the category is slightly better behaved if we allow not necessarily unital algebras and homomorphisms. That is, $\mathbf{Alg}_R$ is analogous to $\mathbf{Rng}$ rather than $\mathbf{Ring}$.

4.1.4   Modules and representations

In what must seem like a circular fashion, we now want to talk about modules *over algebras*, and to relate these to representations of algebras. The above framework nearly covers what we need but since our algebras may not be commutative, we need to state the definition for clarity and completeness.

Definition 4.1.18. Let $R = (R, m_R, u_R)$ be a commutative unital ring and let $A = (A, m, u)$ be a unital associative algebra over R.

Let M be an Abelian group. We say that the pair $(M, \triangleright)$ is a *left A-module* if $\triangleright\colon A \otimes_R M \to M$ is a homomorphism of R-modules such that the diagrams

$$
\begin{array}{ccc}
A \otimes_R A \otimes_R M & \xrightarrow{\;m \otimes \mathrm{id}\;} & A \otimes_R M \\
{\scriptstyle \mathrm{id} \otimes \triangleright} \downarrow & & \downarrow {\scriptstyle \triangleright} \\
A \otimes_R M & \xrightarrow[\;\triangleright\;]{} & M
\end{array}
$$

and

$$
\begin{array}{ccc}
R \otimes_R M & \xrightarrow{\;u_R \otimes \mathrm{id}\;} & A \otimes_R M \\
& \searrow \qquad \swarrow {\scriptstyle \triangleright} & \\
& M &
\end{array}
$$

commute.

The conditions in this definition are equivalent to:

(a) $(a + b) \triangleright m = (a \triangleright m) + (b \triangleright m)$ for all $a, b \in A$, $m \in M$;

(b) $a \triangleright (m + n) = (a \triangleright m) + (a \triangleright n)$ for all $a \in A$, $m, n \in M$;

(c) $a \triangleright (b \triangleright m) = ab \triangleright m$ for all $a, b \in R$, $m \in M$; and

(d) $1_A \triangleright m = m$ for all $m \in M$.

We can define homomorphisms entirely analogously.

Definition 4.1.19. Let $R = (R, m_R, u_R)$ be a commutative unital ring and let $A = (A, m, u)$ be a unital associative algebra over R.

Let $(M, \triangleright_M)$, $(N, \triangleright_N)$ be left A-modules. We say that a map $f\colon M \to N$ is an *A-module homomorphism* if f is a homomorphism of R-modules and the following diagram commutes:

$$
\begin{array}{ccc}
A \otimes_R M & \xrightarrow{\;\mathrm{id} \otimes f\;} & A \otimes_R N \\
{\scriptstyle \triangleright_M} \downarrow & & \downarrow {\scriptstyle \triangleright_N} \\
M & \xrightarrow[\;f\;]{} & N
\end{array}
$$

That is, if $f(a \triangleright_M m) = a \triangleright_N f(m)$ for all $a \in A$ and $m \in M$.

Then there is a category A-Mod, the category of left A-modules, whose objects are left A-modules and morphisms A-module homomorphisms. When A is an R-algebra, there is a forgetful functor $\mathcal{F}\colon A\text{-Mod} \to R\text{-Mod}$. For if $(M, \triangleright)$ is an A-module, forgetting the A-action $\triangleright$ leaves us with M, which is assumed to be an R-module from the start. Also, an A-module homomorphism is required to be an R-module homomorphism.

As previously, we have some general examples—recorded here mainly for completeness—but the seriously interesting examples will have to wait for Chapter 5, after we have laid the groundwork with the theory of modules for algebras in the rest of this chapter.

Example 4.1.20. The *zero module* is the module $(\{0\}, \triangleright = 0)$ with $a \triangleright 0 = 0$ for all $a \in A$. We will usually save ourselves some writing by calling this module 0 rather than $\{0\}$. Any other (sub)module is called *non-zero*.

Example 4.1.21. Again, we always have $_A A = (A, m, u)$, the *regular* (left) A-module. (Compare with Example 4.1.12 above.)

As with other algebraic structures, we have the notions of the kernel and image of a module homomorphism.

Definition 4.1.22. Let $f \colon M \to N$ be a homomorphism of A-modules. The kernel of f is defined to be the subset of M given by

$$\operatorname{Ker} f = \{m \in M \mid f(m) = 0\}.$$

We also denote by $\operatorname{Im} f$ the usual image (as a function) of f.

Now, to tie together the previous chapter and this one, we show that representations of a $\mathbb{K}$-algebra A and A-modules are "the same thing". For the rest of this section, we will work over a field $\mathbb{K}$; the interested reader can examine the constructions and modify as needed for the more general case of R a commutative unital ring.

Proposition 4.1.23. *Let $A = (A, m, u)$ be a $\mathbb{K}$-algebra and M a $\mathbb{K}$-vector space.*

  (i) *An A-module structure $(M, \triangleright)$ on M gives rise to a canonical representation of A on M.*

 (ii) *A representation ρ of A on M gives rise to a canonical A-module structure on M.*

(iii) *The constructions in (i) and (ii) are mutually inverse to each other.*

Proof. We use the much more general technical result known as Hom-$\otimes$ *adjunction*, covered in Section 2.7. In the case we are interested in, this says that the function

$$\varphi \colon \operatorname{Hom}_{\mathbf{Vect}_{\mathbb{K}}}(A \otimes M, M) \to \operatorname{Hom}_{\mathbf{Vect}_{\mathbb{K}}}(A, \operatorname{Hom}_{\mathbf{Vect}_{\mathbb{K}}}(M, M))$$

given by

$$\triangleright \mapsto \rho_{\triangleright} \overset{\text{def}}{=} (a \mapsto L_a)$$

is an isomorphism, where $L_a \colon M \to M$ is the map $L_a(m) = a \triangleright m$. (We have used the same name, L_a, as for the "left multiplication by a" map we have seen for groups before; this is appropriate because the former generalizes the latter.) Note too that we have written the "long" name, $\operatorname{Hom}_{\mathbf{Vect}_{\mathbb{K}}}(M, M)$, for $\operatorname{End}_{\mathbb{K}}(M)$, so that elements of the right-hand side are indeed linear maps $A \to \operatorname{End}_{\mathbb{K}}(M)$, i.e. could be representations, if they have the right properties.

Elements of the left-hand side are linear maps $A \otimes M \to M$: we claim that if $\triangleright \colon A \otimes M \to M$ gives an A-module structure, then $\rho_\triangleright$ will actually be a representation.

To check this, we track through the definitions. For example, $\rho_\triangleright(ab) = L_{ab}$ so

$$\rho_\triangleright(ab)(m) = L_{ab}(m) = (ab)\triangleright m = a\triangleright(b\triangleright m) = (L_a \circ L_b)(m) = (\rho_\triangleright(a) \circ \rho_\triangleright(b))(m)$$

as required.

For the converse direction, let $\rho \in \mathrm{Hom}_{\mathbf{Vect}_\mathbb{K}}(A, \mathrm{Hom}_{\mathbf{Vect}_\mathbb{K}}(M, M))$. Then define $\triangleright_\rho \colon A \otimes M \to M$ by $\triangleright_\rho(a \otimes m) = \rho(a)(m)$. Here, $\rho(a)$ is a map $M \to M$, so we can apply this to m to get an element of M. Similarly to above,

$$a \triangleright_\rho (b \triangleright_\rho m) = a \triangleright_\rho (\rho(b)(m)) = (\rho(a) \circ \rho(b))(m) = \rho(ab)(m) = ab \triangleright_\rho m$$

so that if ρ is a representation, then $\triangleright_\rho$ becomes a left action.

It is straightforward to check (so do it!) that $\rho \mapsto \triangleright_\rho$ is inverse to φ (which sends $\triangleright$ to $\rho_\triangleright$). □

Definition 4.1.24. Let $A = (A, m, u)$ be a $\mathbb{K}$-algebra and let ρ and ρ' be representations of A on V and V' respectively. A *homomorphism* of representations is a $\mathbb{K}$-linear map $f \colon V \to V'$ such that for all $a \in A$, $\rho'(a) \circ f = f \circ \rho(a)$. Denote the set of homomorphisms from ρ to ρ' by $\mathrm{Hom}_A(\rho, \rho')$.

Let $\mathrm{Rep}(A)$ be the category with objects representations of A and morphisms homomorphisms of representations.

Theorem 4.1.25. *The categories* $\mathrm{Rep}(A)$ *and* A*-*Mod *are equivalent.*

Proof. We claim that there are functors

$$F \colon A\text{-}\mathrm{Mod} \to \mathrm{Rep}(A), \ F(M, \triangleright) = \varphi(\triangleright), \ Ff = f$$

and

$$G \colon \mathrm{Rep}(A) \to A\text{-}\mathrm{Mod}, \ G\rho = (M, \varphi^{-1}(\rho)), \ Gf = f$$

where $\rho \colon A \to \mathrm{End}_\mathbb{K}(M)$.

Proposition 4.1.23 asserts that $FG = I_{\mathrm{Rep}(A)}$ and $GF = I_{A\text{-}\mathrm{Mod}}$ on objects and this is clearly the case on morphisms. □

It is worth unpacking the perhaps odd claim that $Ff = f$ and $Gf = f$ are valid choices. If $f \colon M \to N$ is an A-module homomorphism, then $f \colon M \to N$ is a $\mathbb{K}$-linear map such that $f \circ \triangleright_M = \triangleright_N \circ (\mathrm{id} \otimes f)$. We should check that $\rho_{\rhd_N} \circ f = f \circ \rho_{\rhd_M}(a)$. Indeed, we have

$$(\rho_{\rhd_N}(a) \circ f)(m) = \rho_{\rhd_N}(a)(f(m))$$
$$a \rhd_N f(m)$$
$$= \rhd_N(a \otimes f(m))$$
$$= (\rhd_N \circ (\mathrm{id} \otimes f))(a \otimes m)$$
$$\overset{(*)}{=} (f \circ \rhd_M)(a \otimes m)$$
$$= f(a \rhd_M m)$$
$$= f(\rho_{\rhd_M}(a)(m))$$
$$= (f \circ \rho_{\rhd_M}(a))(m)$$

as required. The check for $Gf = f$ (showing that the homomorphism of representations condition translates to the homomorphism of modules condition) is achieved by breaking the sequence of equalities at $(*)$ and noting that then the assumption is the equality of the two ends. That is, one turns the proof inside out!

In fact we have shown *isomorphism* of the categories—it is unusual one can say something so strong.

This is the mathematical sense in which representations of A are the same as A-modules—but even more, homomorphisms of representations are the same as homomorphisms of modules.

4.2   Submodules, quotients and extensions

In this section, $A = (A, m_A, u_A)$ will again be an R-algebra over R a commutative unital ring.

Recall from the previous section that $M = (M, \rhd)$ is a left A-module if M is an R-module and $\rhd \colon A \otimes M \to M$ is such that $\rhd \circ (m_A \otimes \mathrm{id}) = \rhd \circ (\mathrm{id} \otimes \rhd)$ and $\rhd \circ (u_A \otimes \mathrm{id}) = s$.

We will take a categorical approach and express as many concepts in terms of homomorphisms of modules as we can. So recall too that if $(M, \rhd_M)$ and $(N, \rhd_N)$ are A-modules, $f \colon M \to N$ is a homomorphism of modules if f is an R-module map such that $f \circ \rhd_M = \rhd_N \circ (\mathrm{id} \otimes f)$. An *isomorphism* of modules is a module homomorphism admitting an inverse module homomorphism; this is equivalent to the homomorphism being bijective.

We start with the notion of a submodule.

Definition 4.2.1. Let $M = (M, \rhd)$ be an A-module. Let N be a subspace of M such that $\rhd(A \otimes N) \leq N$ (that is, for all $a \in A, n \in N$, we have $a \rhd n \in N$). Then $(N, \rhd|_{A \otimes N})$ is called a *submodule* of M.

We will mildly abuse notation by writing $\rhd$ for $\rhd|_{A \otimes N}$.

One should check that $(N, \triangleright)$ is an A-module, i.e. that the required properties for $\triangleright|_{A \otimes N}$ follow from them being satisfied by $\triangleright$; essentially, we can just observe that the conditions can be checked element-wise.

Example 4.2.2. Every non-zero module M has at least two submodules, 0 and M itself. (The zero module has just one, of course.) Any submodule of M not equal to M is called *proper*.

Note that under the correspondence between representations and modules, subrepresentations correspond to submodules.

The definition we gave of a submodule is "not very categorical": if N and N' are isomorphic but not equal, N being a submodule of M does not imply that N' is. The reason is that N is actually a subset of M, whereas we only know that N' is in bijection with one. Sometimes we really care about actual subobjects but within A-Mod, we should relax this and consider morphisms that are *monic*. In A-Mod, this is equivalent to being an injective homomorphism.

Lemma 4.2.3. *Let $\iota\colon N \to M$ be an injective A-module homomorphism. Then N is isomorphic to a submodule of M, namely $\operatorname{Im} \iota$.*

Conversely, if N is a submodule of M, there is an injective A-module homomorphism $\iota\colon N \to M$ given by $\iota(n) = n$.

Indeed, for completeness, let us record that kernels and images for module homomorphisms behave as we would expect (especially remembering that if $R = \mathbb{K}$ we just have vector spaces and 'submodule' precisely means 'subspace').

Proposition 4.2.4. *Let $f\colon M \to N$ be a homomorphism of A-modules.*

  (i) *The kernel of f, $\operatorname{Ker} f$, is a submodule of M.*

 (ii) *The image of f, $\operatorname{Im} f$, is a submodule of N.*

(iii) *The homomorphism f is injective if and only if $\operatorname{Ker} f = 0$.*

(iv) *The homomorphism f is surjective if and only if $\operatorname{Im} f = N$.*

Proof. Exercise. $\qquad\square$

We said that quotient representations were awkward to define. One reason for studying modules instead is that quotients are much easier to treat, as follows.

Definition 4.2.5. Let $M = (M, \triangleright)$ be an A-module and $N = (N, \triangleright)$ a submodule of M. The *quotient module* M/N has underlying Abelian group M/N and module structure map $\bar{\triangleright}\colon A \otimes (M/N) \to M/N$ given by $\bar{\triangleright}(a \otimes (m+N)) = (a \triangleright m) + N$.

More explicitly, the action of $a \in A$ on a coset $m + N \in M/N$ is

$$a \,\bar{\triangleright}\, (m + N) = (a \triangleright m) + N.$$

Notice that we do not need to ask for anything extra on N—just that it is a submodule—unlike groups or rings, where we need a normal subgroup and an ideal respectively. This is mainly due to the fact that any subgroup of an Abelian group is normal, and so quotients are defined for any submodule of a module.

We show the key module property for $\bar{\triangleright}$:

$$
\begin{aligned}
ab\,\bar{\triangleright}\,(m + N) &= (ab \triangleright m) + N \\
&= (a \triangleright (b \triangleright m)) + N \\
&= a\,\bar{\triangleright}\,((b \triangleright m) + N) \\
&= a\,\bar{\triangleright}\,(b\,\bar{\triangleright}\,(m + N)).
\end{aligned}
$$

Note too that our definition is being a little slick: the role of the R-module structure is hidden. Due to the compatibilities we ask for, the above is not incorrect, but arguably it would have been better to say "the quotient A-module has underlying R-module M/N and A-module structure map $\bar{\triangleright}\colon A \otimes (M/N) \to M/N$". For this, one essentially does the previous verification twice: once with R instead of A, to see that M/N inherits an R-module structure and then as above, to get the A-module structure.

As with submodules, being a quotient is not invariant under isomorphism. This time, the "right" generalization is to consider *epis*, which in A-Mod is equivalent to being a surjective homomorphism.

Lemma 4.2.6. *Let $\pi\colon M \to N$ be a surjective A-module homomorphism. Then N is isomorphic to a quotient module of M, namely $M/\operatorname{Ker}\pi$.*

Conversely, if M/N is a quotient module of M by a submodule N, there is a surjective A-module homomorphism $\pi\colon M \to M/N$ given by $\pi(m) = m + N$.

Since $\operatorname{Ker}\pi$ is a submodule of M (by the proposition-exercise), there is an associated injective homomorphism ι so that we could write

$$
\operatorname{Ker}\pi \overset{\iota}{\hookrightarrow} M \overset{\pi}{\twoheadrightarrow} N
$$

At M, we have $\pi \circ \iota = 0$ (everything in $\operatorname{Im}\iota = \operatorname{Ker}\pi$ is sent to 0 by π). Now ι being injective means that if we write $0\colon 0 \to \operatorname{Ker}\pi$ for the (unique) map sending 0 to 0, we have $\operatorname{Ker}\iota = 0 = \operatorname{Im}0$. Similarly, writing $0\colon N \to 0$ for the (unique) map defined by $0(n) = 0$ for all $n \in N$, we have $\operatorname{Im}\pi = N = \operatorname{Ker}0$; this is equivalent to π being surjective. We usually suppress the labels for zero maps and write all of this as

$$
0 \longrightarrow \operatorname{Ker}\pi \overset{\iota}{\hookrightarrow} M \overset{\pi}{\twoheadrightarrow} N \longrightarrow 0
$$

noting that at each object X, $\operatorname{Im}(\to X) = \operatorname{Ker}(X \to)$. We call this equality of the image of the incoming map with the kernel of the outgoing map *exactness* at X. Replacing $\operatorname{Ker}\pi$ with anything isomorphic to it, we obtain the following definition.

Definition 4.2.7. Let K, M and N be A-modules.

If $f \colon K \to M$ and $g \colon M \to N$ are module homomorphisms such that f is injective, g is surjective and $\operatorname{Im} f = \operatorname{Ker} g$, then we say that

$$0 \longrightarrow K \overset{f}{\hookrightarrow} M \overset{g}{\twoheadrightarrow} N \longrightarrow 0$$

is a short exact sequence of A-modules.

It is called "short" because, well, there are long ones too. Short exact sequences are sometimes called *extensions*, because they describe a way to "extend" K by N to form a bigger module M, having a submodule isomorphic to K with quotient isomorphic to N.

Some short exact sequences are special.

Definition 4.2.8. A short exact sequence

$$0 \longrightarrow K \overset{f}{\hookrightarrow} M \overset{g}{\twoheadrightarrow} N \longrightarrow 0$$

of A-modules is *split* if there exists an A-module homomorphism $u \colon N \to M$ such that $g \circ u = \operatorname{id}_N$.

Note that u is necessarily injective. In nice categories—the correct name is *Abelian categories* (see Section 4.2.1 below)—such as A-Mod, split short exact sequences arise in a particularly natural way.

Definition 4.2.9. Let $M = (M, \triangleright_M)$ and $N = (N, \triangleright_N)$ be A-modules. Then the Cartesian product $M \times N$ is given an A-module structure via

$$\triangleright_\oplus \colon A \otimes (M \times N) \to M \times N, \ \triangleright_\oplus(a \otimes (m, n)) = (a \triangleright_M m, a \triangleright_N n).$$

Then $(M \times N, \triangleright_\oplus)$ is called the *direct sum* of M and N and we write $M \oplus N$.

In the case $R = \mathbb{K}$, this is exactly the vector space direct sum $M \oplus N$ made into an A-module in the natural way: $a \triangleright (m + n) = a \triangleright m + a \triangleright n$.

We have two canonical injective maps $\iota_M \colon M \to M \oplus N$, $\iota_M(m) = m + 0$ and $\iota_N \colon N \to M \oplus N$, $\iota_N(n) = 0 + n$, and two canonical surjective maps $\pi_M \colon M \oplus N \to M$, $\pi_M(m + n) = m$ and $\pi_N \colon M \oplus N \to N$, $\pi_N(m + n) = n$.

Proposition 4.2.10. *The following are equivalent:*

(i) $0 \longrightarrow K \overset{f}{\hookrightarrow} M \overset{g}{\twoheadrightarrow} N \longrightarrow 0$ *is a split short exact sequence of A-modules;*

(ii) $0 \longrightarrow K \overset{f}{\hookrightarrow} M \overset{g}{\twoheadrightarrow} N \longrightarrow 0$ *is a short exact sequence of A-modules and there exists a module homomorphism $t \colon M \to K$ such that $t \circ f = \operatorname{id}_K$ (such a map t is necessarily surjective); and*

(iii) *there is an isomorphism* $h\colon M \to K \oplus N$ *such that* $h \circ f = \iota_K$ *is the canonical injective map and* $g \circ h^{-1} = \pi_N$ *is the canonical surjective map.*

Proof. Exercise. Hint: for (iii) implies (i), consider

$$
\begin{array}{ccccccccc}
0 & \longrightarrow & K & \overset{f}{\longrightarrow} & M & \overset{g}{\twoheadrightarrow} & N & \longrightarrow & 0 \\
 & & \| & & \downarrow{\scriptstyle h} & & \| & & \\
0 & \longrightarrow & K & \overset{\iota_K}{\underset{\pi_K}{\rightleftarrows}} & K \oplus N & \overset{\iota_N}{\underset{\pi_N}{\rightleftarrows}} & N & \longrightarrow & 0
\end{array}
$$

Then show that setting $u = h^{-1} \circ \iota_N$ we have a map showing that the sequence splits. $\qquad\square$

This is the starting point for the construction of an object that classifies extensions of modules. The details would require too much time and technicalities beyond us at this point, but the idea is accessible: take all short exact sequences starting with K and ending with N, define an equivalence relation on these and hence construct a group $\mathrm{Ext}^1(K, N)$, in which $[0 \to K \to K \oplus N \to N \to 0]$ is the identity element. In particular, $\mathrm{Ext}^1(K, N) = \{0\}$ if and only if every extension of K by N is split (i.e. the only extension is the trivial one, $K \oplus N$).

In some sense, one major goal of representation theory is to understand $\mathrm{Hom}(K, N)$, $\mathrm{Ext}^1(K, N)$ and the higher extension groups $\mathrm{Ext}^i(K, N)$ for all modules K, N for some given algebra. In fact, there is usually a subsidiary goal before this, namely to understand all of the "fundamental pieces" from which we can build other modules via extensions. This is what we turn our attention to in the next section.

4.2.1 ♟ Abelian categories

The definition of an Abelian category is designed to capture some key properties of the category **Ab** of Abelian groups, or more generally (recalling that **Ab** $= \mathbb{Z}$-Mod) the categories R-Mod. As a result, many things one might like to do in a category of modules (and such "things" might include *homological algebra*) can be done in a general Abelian category.

The definition has several pieces, as follows:

- the category $\mathcal{A}$ should be **Ab**-enriched, meaning that every morphism set $\mathrm{Hom}_{\mathcal{A}}(A, B)$ is an Abelian group (in this way, we can add morphisms and there is a zero morphism) and that the maps induced by composition are morphisms in **Ab**;

- $\mathcal{A}$ has a zero object (an object 0 such that every other object A has a unique morphism $A \to 0$ and a unique morphism $0 \to A$);

- $\mathcal{A}$ should have all finite products and coproducts (with the preceding two conditions, such a category is called *additive*);

- $\mathcal{A}$ should admit (arbitrary) kernels and cokernels; and

- every monomorphism is a kernel and every epimorphism is a cokernel.

The category **Ab** is indeed Abelian. If $f, g \in \mathrm{Hom}_{\mathbf{Ab}}(A, B)$, defining $(f+g)(a) = f(a)+g(a)$ for all $a \in A$ gives us the means to add homomorphisms and so morphism sets are Abelian groups. The zero object is the trivial group $\{e\}$: given an Abelian group A, $\{e\}$ is a subgroup in a unique way and conversely there is always a surjective group homomorphism $\pi \colon A \twoheadrightarrow \{e\}$ given by $\pi(a) = e$. We can form both finite Cartesian products and finite direct sums of Abelian groups. In **Ab**, arbitrary kernels and cokernels exist, monomorphisms are injective and their kernels exist (the usual $\mathrm{Ker}\, f$) and similarly epimorphisms are surjective and their cokernels exist (by taking $N/\mathrm{Im}\, f$).

This extends to module categories too, with appropriate replacements, e.g. the zero module for the trivial group. Recall that this includes $\mathbf{Vect}_{\mathbb{K}} = \mathbb{K}\text{-Mod}$.

Although not every Abelian category is a category of modules over a ring, things are not so bad. By the Freyd-Mitchell embedding theorem (see e.g. [Wei, Theorem 1.6.1]), every Abelian category $\mathcal{A}$ is a full subcategory of a category of modules over some ring R such that the embedding functor $\mathcal{A} \to R\text{-Mod}$ is an exact functor. Furthermore, if $\mathcal{A}$ has all small coproducts and a compact projective generator, then $\mathcal{A}$ is equivalent to $R\text{-Mod}$ for some ring R.

Although far beyond the scope of this book, we mention one last class of examples, to show that these ideas transcend representation theory in the sense usually meant by that phrase at undergraduate level. Namely, the category of sheaves of Abelian groups on a topological space (or indeed, on still more general "spaces") is Abelian.

4.2.2   Noether's isomorphism theorems for modules

We will on occasion need to refer to the various isomorphism theorems for modules. We will state these without proof and just note that to prove them, one observes that the corresponding results for groups apply (since modules are Abelian groups with extra structure) and so what is needed is to check that the various maps constructed are module homomorphisms. One may find this explicitly in e.g. [EH, Theorem 2.24].

We also note that these theorems are due to Emmy Noether.

Problem 22. Find out about Emmy Noether and tell someone about her.

Theorem 4.2.11. *Let A be an R-algebra for R a commutative unital ring.*

Universal property of quotients

Let M and P be A-modules. Let N be a submodule of M and let $\pi\colon M \to M/N$ be the associated quotient homomorphism.

Then for every A-module homomorphism $\varphi\colon M \to P$ such that $N \subseteq \operatorname{Ker}\varphi$, there exists a unique A-module homomorphism $\bar\varphi\colon M/N \to P$ such that $\varphi = \bar\varphi \circ \pi$.

$$
\begin{array}{ccc}
M & \xrightarrow{\ \varphi\ } & P \\
\ \downarrow{\scriptstyle\pi} & \nearrow{\scriptstyle \exists!\bar\varphi} & \\
M/N & &
\end{array}
$$

Furthermore, $\operatorname{Ker}\bar\varphi = (\operatorname{Ker}\varphi)/N$ and $\operatorname{Im}\bar\varphi = \operatorname{Im}\varphi$.

First Isomorphism Theorem

Let M and P be A-modules and let $\varphi\colon M \to P$ be an A-module homomorphism. Then $M/\operatorname{Ker}\varphi \cong \operatorname{Im}\varphi$.

Second Isomorphism Theorem

Let L and N be submodules of an A-module M. Then their sum $L + N$ and intersection $L \cap N$ are submodules of M and $L/(L \cap N) \cong (L + N)/N$.

Third Isomorphism Theorem

Let $L \leq N \leq M$ be a sequence of submodules. Then N/L is a submodule of M/L and $(M/L)/(N/L) \cong M/N$.

Submodule correspondence

Let N be a submodule of an A-module M and let $\pi\colon M \to M/N$ be the canonical surjection. Then the map $P \mapsto \pi^{-1}(P)$ induces an inclusion-preserving bijection between the submodules of M/N and the submodules of M containing N, whose inverse is given by $L \mapsto M/L$.

4.3   Simple and semisimple modules

Definition 4.3.1. An A-module M is *simple* if it has exactly two submodules, 0 and M.

By definition, the zero module is *not* simple.[4]

---

[4]See the discussion "Too simple to be simple" on the nLab.

Remark 4.3.2 (Important!). Simple modules are also called irreducible, although this term tends to be used more for the corresponding representation. That is, we say a non-zero representation is irreducible if it has only the zero subrepresentation and itself as subrepresentations. Then it is common to shorten "irreducible representation" to "irrep". As we are taking a module-theoretic approach, we will say "simple module".

Example 4.3.3. Let $\mathbb{K}$ be a field and consider the $\mathbb{K}$-algebra $M_n(\mathbb{K})$ of $n \times n$ matrices with entries in $\mathbb{K}$. In a course on linear algebra, one usually proves that this is a $\mathbb{K}$-algebra, although not in so many words. The algebra structure is given by matrix addition, scalar multiplication and matrix multiplication.

The matrix algebra $M_n(\mathbb{K})$ has a natural module, which we (imaginatively) call the *natural module*. Namely, take $\mathbb{K}^n$, thought of as column vectors of length n. This is a left module: $M \triangleright v \stackrel{\text{def}}{=} Mv$, the usual action given by multiplying the vector v by the matrix M on the left.

This module is in fact simple and, up to isomorphism, is the only simple module for $M_n(\mathbb{K})$.

Note for later that all of the above remains true if we replace the field $\mathbb{K}$ by a division ring.

We can combine simples to obtain larger modules:

Definition 4.3.4. A module M is called *semisimple* if it is isomorphic to a (not necessarily finite)[5] direct sum of simple modules.

This notion is also called "completely reducible", usually when the term irreducible is being used, rather than simple. This deserves a name because in general not every module is semisimple—this is not meant to be obvious, but it is very much true.

Lemma 4.3.5. *A left A-module M is semisimple if and only if every submodule of M is a direct summand. Hence, every non-zero submodule and non-zero quotient module of a semisimple module is again semisimple.*

Proof. We will omit this proof, as in full generality (to deal with the "not necessarily finite" cases) it requires some more advanced techniques. Those who are interested will find this result in [Rot, Section 4.1] or [EH, Section 4.1]. □

An algebra A is called semisimple if the regular module $_A A$ is semisimple.

An Abelian category in which every object is semisimple is itself called semisimple.

Simple modules deserve to be considered as building blocks for arbitrary modules for the following reason.

---

[5]An infinite direct sum $\bigoplus_{i \in I} V_i$ has as elements sequences $(v_1, v_2, \dots)$ with $v_i \in V_i$ and all but finitely many v_i zero. The vector space operations are defined component-wise, as for finite direct sums.

Definition 4.3.6. Let M be an A-module. A *composition series* for M is a finite sequence of submodules

$$0 = M_0 \leq M_1 \leq M_2 \leq \cdots \leq M_{r-1} \leq M_r = M$$

such that M_i/M_{i-1} is simple for all $0 \leq i \leq r-1$. We call r the *length* of the composition series. If M has a composition series, we say M has *finite length*.

Note that since the zero module is not simple, the inclusions here are in fact strict, i.e. M_{i-1} is a proper submodule of M_i for any i. That is, there are no repetitions in the series: $0 \leq M \leq M \leq M$ is not allowed (or at least, it is not a composition series of length 3).

We refer to the collection of $\{M_i/M_{i-1} \mid 1 \leq i \leq r\}$ as the set of *subquotients* of the composition series.

Proposition 4.3.7. *Let M be an A-module of finite length. Then every submodule of M has finite length, this being at most the length of M. If N is a proper submodule of M, its length is strictly less than that of M.*

Proof. Let

$$0 = M_0 \leq M_1 \leq M_2 \leq \cdots \leq M_{r-1} \leq M_r = M$$

be a composition series for M. Let N be a submodule of M and define $N_i = N \cap M_i$ for all i. Then

$$0 = N_0 \leq N_1 \leq N_2 \leq \cdots \leq N_{r-1} \leq N_r = N \tag{4.1}$$

is a sequence of submodules, but may *a priori* fail to be a composition series if the subquotients N_i/N_{i-1} are not simple.

Fix i and consider the homomorphism $\varphi_i \colon N_i \to M_i/M_{i-1}$ given by $\varphi_i(n) = n + M_{i-1}$. That is, let φ_i be the composition of the inclusion $N_i = N \cap M_i \hookrightarrow M_i$ with the surjection $M_i \twoheadrightarrow M_i/M_{i-1}$. By construction, the kernel of φ_i is $N_{i-1} = N \cap M_{i-1}$, so by the First Isomorphism Theorem for Modules, there is an injective map $\psi \colon N_i/N_{i-1} \to M_i/M_{i-1}$.

Now M_i/M_{i-1} is simple, since we chose the M_j to give us a composition series for M, so the image of ψ is either zero (in which case $N_i/N_{i-1} = 0$, or equivalently $N_i = N_{i-1}$) or equal to M_i/M_{i-1} (in which case $N_i/N_{i-1} \cong M_i/M_{i-1}$ is simple). Removing any terms from (4.1) where $N_i = N_{i-1}$ we obtain a composition series for N, of length at most r.

We may show by induction on i that if $N_i/N_{i-1} \cong M_i/M_{i-1}$ for all i, then $N_i = M_i$ for all i and in particular $N = M$. Hence if N is a proper submodule, its length must be strictly less than that of M. $\qquad\square$

The fundamental theorem relating to composition series is the following.

Theorem 4.3.8 (Jordan–Hölder). *Let M be an A-module. If M has two composition series*

$$0 \leq M_1 \leq \cdots \leq M_{r-1} \leq M_r = M$$
$$0 \leq N_1 \leq \cdots \leq N_{s-1} \leq N_s = M$$

then $r = s$ and there exists a permutation $\sigma \in S_r$ such that $M_i/M_{i-1} \cong N_{\sigma(i)}/N_{\sigma(i)-1}$ for all i.

Proof. For ease of expression, let us say two composition series of a module M are *equivalent* if they satisfy the hypotheses of the theorem, i.e. they have the same length and there exists a bijection of the sets of subquotients such that each subquotient for the first series is isomorphic to its image under the bijection.

Let

$$0 \leq M_1 \leq \cdots \leq M_{r-1} \leq M_r = M \tag{4.2}$$

and

$$0 \leq N_1 \leq \cdots \leq N_{s-1} \leq N_s = M \tag{4.3}$$

be two composition series of M. We will proceed by induction on r. If $r = 1$ then $M/0 = M$ is simple and $0 \leq M$ is its only composition series. So, let $r > 1$ and assume (for a strong induction) that the theorem holds for any module having some composition series of length less than r.

If $M_{r-1} = N_{s-1}$ then M_{r-1} has two composition series

$$0 \leq M_1 \leq \cdots \leq M_{r-1}$$

and

$$0 \leq N_1 \leq \cdots \leq N_{s-1} = M_{r-1},$$

of length $r-1$ and $s-1$ respectively. By the inductive hypothesis, these series are equivalent, i.e. $r - 1 = s - 1$ and there is a permutation $\sigma \in S_{r-1}$ such that $M_i/M_{i-1} \cong N_{\sigma(i)}/N_{\sigma(i)-1}$ for all $i \leq r-1$. Hence $r = s$ and $M/M_{r-1} \cong M/N_{s-1}$ so, extending σ to $\sigma' \in S_r$ with $\sigma'(r) = r$ and $\sigma'|_{\{1,...,r-1\}} = \sigma$, we have the desired conclusion.

If $M_{r-1} \neq N_{s-1}$, since $M_{r-1} < M$ and $N_{s-1} < M$ are proper submodules, we have $M_{r-1} + N_{s-1} \leq M$. But M/M_{r-1} is simple so we cannot have $M_{r-1} \leq N_{s-1}$, else N_{s-1}/M_{r-1} is a non-trivial proper submodule of M/M_{r-1}. Therefore we must have $N_{s-1} < M_{r-1} + N_{s-1}$, but since M/N_{s-1} is simple, we must have that $M_{r-1} + N_{s-1}$ is equal to M.

Let $K = M_{r-1} \cap N_{s-1} < M$. By the second isomorphism theorem (4.2.11), we have $M/M_{r-1} \cong N_{s-1}/K$ and $M/N_{s-1} \cong M_{r-1}/K$ and so M_{r-1}/K and N_{s-1}/K are simple.

Now by Proposition 4.3.7, K has finite length so has a composition series

$$0 \leq K_1 \leq K_2 \leq \cdots \leq K_{t-1} \leq K_t = K.$$

Then

$$0 \leq M_1 \leq M_2 \leq \cdots \leq M_{r-2} \leq M_{r-1} \tag{4.4}$$

and

$$0 \leq K_1 \leq K_2 \leq \cdots \leq K_{t-1} \leq K \leq M_{r-1} \tag{4.5}$$

are composition series of M_{r-1} of length $r-1$ and $t+1$ respectively. By the inductive hypothesis, these series are equivalent (in particular, $t = r - 2$). Similarly we have composition series

$$0 \leq N_1 \leq N_2 \leq \cdots \leq N_{s-2} \leq N_{s-1} \tag{4.6}$$

and

$$0 \leq K_1 \leq K_2 \leq \cdots \leq K_{t-1} \leq K \leq N_{s-1} \tag{4.7}$$

of N_{s-1} of length $s-1$ and $t+1 = r-1$ respectively, so again by the inductive hypothesis, $r = s$ and (4.6) and (4.7) are equivalent.

Finally, since $M_{r-1}/K \cong M/N_{s-1} = M/N_{r-1}$ and $N_{r-1}/K \cong M/M_{r-1}$, we see that the composition series

$$0 \leq K_1 \leq K_2 \leq \cdots \leq K_{t-1} \leq K \leq M_{r-1} \leq M \tag{4.8}$$

and

$$0 \leq K_1 \leq K_2 \leq \cdots \leq K_{t-1} \leq K \leq N_{s-1} \leq M \tag{4.9}$$

are equivalent. Since (4.4) and (4.5) are equivalent, and (4.6) and (4.7) are equivalent, (4.2) and (4.8) are equivalent and (4.3) and (4.9) are equivalent. Since (4.8) and (4.9) are equivalent, (4.2) and (4.3) are equivalent.[6]   □

Since then any two composition series for a module M have the same length, this implies that if M has a composition series (and it may not!) then M has well-defined length, the length of any composition series for it.

Now, if M has a composition series $0 \leq M_1 \leq \cdots \leq M_{r-1} \leq M_r = M$, we have that

- M_1 is simple;

- $0 \to M_1 \to M_2 \to M_2/M_1 \to 0$ is a short exact sequence with M_2/M_1 also simple;

- $0 \to M_2 \to M_3 \to M_3/M_2 \to 0$ is a short exact sequence with M_3/M_2 also simple;

- $\ldots$

- $0 \to M_{r-1} \to M_r = M \to M/M_{r-1} \to 0$ is a short exact sequence with M/M_{r-1} also simple.

---

[6]We recommend drawing some diagrams to help understand these relationships.

In other words, M is an iterated extension by simple modules.

In the case that $R = \mathbb{K}$ is a field and A is a $\mathbb{K}$-algebra, we can use the fact that dimension is a well-behaved invariant to show that lots of A-modules do have composition series.

Lemma 4.3.9. *Let A be a $\mathbb{K}$-algebra for $\mathbb{K}$ a field. Every finite-dimensional A-module has a composition series.*

Proof (sketch). Work by (strong) induction on dimension. If $\dim M = 0$, there is nothing to do. If $\dim M > 1$ and M is simple, $0 \leq M$ is a composition series ($M/0 \cong M$).

Otherwise let N be a maximal proper submodule of M (i.e. $N < M$ and if there exists P such that $N \leq P \leq M$ then $P = N$ or $P = M$); finite-dimensionality of M ensures such an N exists. Then M/N is simple. But $\dim N < \dim M$ so by the inductive hypothesis N has a composition series $0 \leq N_1 \leq \cdots \leq N_r = N$, meaning that $0 \leq N_1 \leq \cdots \leq N_r = N < M$ is a composition series for M. $\qquad\square$

At this point, we recommend taking a break from reading and trying the following exercise, parts of which have been mentioned or hinted at above.

Problem 23. Let U, V, W be A-modules for an R-algebra A, and let $\varphi_1, \varphi_2 \in \mathrm{Hom}_{A\text{-Mod}}(U, V)$ and $\psi \in \mathrm{Hom}_{A\text{-Mod}}(V, W)$.

(a) Prove that the composition $\psi \circ \varphi_1 : U \to W$ is an A-module homomorphism.

(b) Prove that the function $(\varphi_1 + \varphi_2) : U \to V$ defined by

$$(\varphi_1 + \varphi_2)(u) = \varphi_1(u) + \varphi_2(u)$$

for all $u \in U$ makes $\mathrm{Hom}_{A\text{-Mod}}(U, V)$ into an Abelian group.

(c) Defining $r \triangleright \varphi_1 : U \to V$, $(r \triangleright \varphi_1)(u) = r \triangleright \varphi_1(u)$ for $r \in R$, show that $\mathrm{Hom}_{A\text{-Mod}}(U, V)$ is an R-module.

(d) Suppose that $U = V$. Prove that $\mathrm{End}_A(U) \overset{\mathrm{def}}{=} \mathrm{Hom}_{A\text{-Mod}}(U, U)$ is an algebra for the addition and R-module structure defined above and with the multiplication being the composition of A-homomorphisms. Give the multiplicative and the additive identity elements.

(e) Is $\mathrm{End}_A(U)$ commutative as a ring?

The following is a foundational result—it is called a lemma only because its proof is relatively straightforward, rather than because it is insignificant. Recall that a *division ring* is a ring such that $R^{\times} = R \setminus \{0\}$, i.e. one in which every non-zero element has a multiplicative inverse. If A is an R-algebra, we say A is a *division algebra* if as a ring, it is a division ring (i.e. $A^{\times} = A \setminus \{0\}$).

Lemma 4.3.10 (Schur's lemma). *Let S be a simple A-module and M an A-module. Then every non-zero homomorphism in $\mathrm{Hom}_{A\text{-Mod}}(S, M)$ is injective and every non-zero homomorphism in $\mathrm{Hom}_{A\text{-Mod}}(M, S)$ is surjective.*

Hence $\mathrm{End}_A(S) \overset{\text{def}}{=} \mathrm{Hom}_{A\text{-Mod}}(S, S)$ is a division R-algebra.

Proof. If $f \in \mathrm{Hom}_{A\text{-Mod}}(S, M)$ and $f \neq 0$, then $\mathrm{Ker}\, f \leq S$ and $\mathrm{Ker}\, f \neq S$ so, as S is simple, we must have $\mathrm{Ker}\, f = 0$ and f is injective.

Similarly, if $g \in \mathrm{Hom}_{A\text{-Mod}}(M, S)$ and $g \neq 0$, then $\mathrm{Im}\, g \leq S$ and $\mathrm{Im}\, g \neq 0$ so, again as S is simple, we must have $\mathrm{Im}\, g = S$ and g is surjective.

Then any non-zero map in $\mathrm{End}_A(S) \overset{\text{def}}{=} \mathrm{Hom}_{A\text{-Mod}}(S, S)$ is both injective and surjective and hence is an isomorphism. That is, every non-zero element of $\mathrm{End}_A(S)$ is invertible, i.e. $\mathrm{End}_A(S)$ is a division algebra. $\qquad\square$

The next closely-related result is also sometimes called "Schur's lemma": its assumptions are stronger but its conclusion is also stronger. It is "morally" a corollary of Schur's lemma, but our formulation makes this a little less clear and our proof is independent of Schur's lemma, so we just refer to it as an extra lemma.

The proof uses the fact that if A is a $\mathbb{K}$-algebra for some field $\mathbb{K}$, then $\mathrm{Hom}_{A\text{-Mod}}(M, N)$ is a $\mathbb{K}$-vector space (cf. Problem 23).

Lemma 4.3.11. *Let $\mathbb{K}$ be a field and A a $\mathbb{K}$-algebra.*

If $\mathbb{K}$ is algebraically closed and S is a finite-dimensional simple module, then $\mathrm{End}_A(S) \cong \mathbb{K}$.

In particular, any homomorphism $f \colon S \to S$ is a scalar multiple of the identity, i.e. $f = \lambda\,\mathrm{id}_S$ for some $\lambda \in \mathbb{K}$.

Proof. Let $f \in \mathrm{End}_A(S)$. Then since $\mathbb{K}$ is algebraically closed and S is finite-dimensional, there exists $0 \neq s \in S$ and $\lambda \in \mathbb{K}$ such that $f(s) = \lambda s$ (i.e. f has an eigenvalue on S; this is what algebraic closure of $\mathbb{K}$ does for us). Since $\lambda\,\mathrm{id}_S$ is also an A-module homomorphism, so is $f - \lambda\,\mathrm{id}_S$.

Now $\mathrm{Ker}(f - \lambda\,\mathrm{id}_S) \leq S$ and $s \in \mathrm{Ker}(f - \lambda\,\mathrm{id}_S)$, so this kernel is non-zero. Since S is simple, $\mathrm{Ker}(f - \lambda\,\mathrm{id}_S) = S$, i.e. $f = \lambda\,\mathrm{id}_S$ on all of S.

Since $\lambda\,\mathrm{id}_S \in \mathrm{End}_A(S)$ for all $\lambda \in \mathbb{K}$, $\{\lambda\,\mathrm{id}_S \mid \lambda \in \mathbb{K}\} \subseteq \mathrm{End}_A(S)$. Conversely we have shown above that $\mathrm{End}_A(S) \subseteq \{\lambda\,\mathrm{id}_S \mid \lambda \in \mathbb{K}\}$. Hence these vector spaces are equal; note finally that $\{\lambda\,\mathrm{id}_S \mid \lambda \in \mathbb{K}\} \cong \mathbb{K}$ as algebras. $\quad\square$

To understand what is going on here, as applications of Schur's lemma and 4.3.11, we recommend the following exercises. Although we have included solutions, as so often, one learns more by having a go first oneself.

Problem 24. Let A be an algebra over $\mathbb{C}$, recalling that $\mathbb{C}$ is an algebraically closed field. Let S and T be non-isomorphic finite-dimensional simple A-modules.

  (a) Determine $\dim \mathrm{Hom}_{A\text{-Mod}}(S, S)$.

  (b) Using Schur's lemma, determine $\dim \mathrm{Hom}_{A\text{-Mod}}(S, T)$.

(c) One can show that $\mathrm{Hom}_{A\text{-Mod}}(-,-)$ is bi-additive, that is, for any A-modules M, N and P, we have

$$\mathrm{Hom}_{A\text{-Mod}}(M \oplus N, P) \cong \mathrm{Hom}_{A\text{-Mod}}(M, P) \oplus \mathrm{Hom}_{A\text{-Mod}}(N, P)$$

and

$$\mathrm{Hom}_{A\text{-Mod}}(M, N \oplus P) \cong \mathrm{Hom}_{A\text{-Mod}}(M, N) \oplus \mathrm{Hom}_{A\text{-Mod}}(M, P).$$

Let $M = S \oplus T$. Using these facts, find the dimension of $\mathrm{End}_A(M) \overset{\text{def}}{=} \mathrm{Hom}_{A\text{-Mod}}(M, M)$, justifying your answer.

Solution.

(a) By Lemma 4.3.11, $\mathrm{End}_A(S) = \mathrm{Hom}_{A\text{-Mod}}(S, S) \cong \mathbb{C}$. So we have that $\dim \mathrm{Hom}_{A\text{-Mod}}(S, S) = 1$.

(b) Consider $f \in \mathrm{Hom}_{A\text{-Mod}}(S, T)$ and assume $f \neq 0$. Since S is simple, by Schur's lemma, f is injective. Similarly, since T is simple, f is also surjective. So f is an isomorphism between S and T, but we assumed that S and T were not isomorphic. So there is a contradiction and we conclude that f must be the zero homomorphism. So $\mathrm{Hom}_{A\text{-Mod}}(S, T) = 0$ and in particular $\dim \mathrm{Hom}_{A\text{-Mod}}(S, T) = 0$.

(c) From the stated facts,

$$\begin{aligned}
\mathrm{End}_A(M) &= \mathrm{Hom}_{A\text{-Mod}}(M, M) \\
&= \mathrm{Hom}_{A\text{-Mod}}(S \oplus T, S \oplus T) \\
&\cong \mathrm{Hom}_{A\text{-Mod}}(S, S \oplus T) \oplus \mathrm{Hom}_{A\text{-Mod}}(T, S \oplus T) \\
&\cong \mathrm{Hom}_{A\text{-Mod}}(S, S) \oplus \mathrm{Hom}_{A\text{-Mod}}(S, T) \oplus \mathrm{Hom}_{A\text{-Mod}}(T, S) \\
&\quad \oplus \mathrm{Hom}_{A\text{-Mod}}(T, T)
\end{aligned}$$

But by part(a), $\dim \mathrm{Hom}_{A\text{-Mod}}(S, S) = \dim \mathrm{Hom}_{A\text{-Mod}}(T, T) = 1$ and by part (b), $\dim \mathrm{Hom}_{A\text{-Mod}}(S, T) = \dim \mathrm{Hom}_{A\text{-Mod}}(T, S) = 0$.

So, $\dim \mathrm{End}_A(M) = 2$.

Note that a similar example to the algebra appearing in the next question was introduced in Problem 7.

Problem 25. Let $A = \mathbb{C}[x]/(x^4 - 1)$.

(a) Let M be the 3-dimensional vector space with basis $\mathcal{B}_M = \{m_1, m_2, m_3\}$ and define

$$\begin{aligned}
x \triangleright m_1 &= -m_1 + (1 + i)m_2 \\
x \triangleright m_2 &= im_2 \\
x \triangleright m_3 &= (1 - i)m_1 + (-1 + 3i)m_2 - im_3
\end{aligned}$$

Show that $\triangleright$ extends to a map $\triangleright \colon A \otimes M \to M$ so that M becomes an A-module.

(b) Decompose M into a direct sum of simple submodules.

(c) Using Schur's lemma and related results, find a basis for $\operatorname{End}_A(M)$.

Solution.

(a) We write down the matrix X corresponding to the action of x:

$$X = \begin{pmatrix} -1 & 0 & 1-i \\ 1+i & i & -1+3i \\ 0 & 0 & -i \end{pmatrix}$$

Then

$$X^2 = \begin{pmatrix} 1 & 0 & -2 \\ -2 & -1 & 2 \\ 0 & 0 & -1 \end{pmatrix}$$

and hence $X^4 = I$ as required for this to define an action of A.

(b) The characteristic polynomial of X is $t^3 + t^2 + t + 1$ so its eigenvalues are -1, i and $-i$. Define S_{-1}, S_i and S_{-i} to be the eigenspaces for the action of X on M. Since these are 1-dimensional, they are simple and since eigenspaces for distinct eigenvalues intersect trivially, we have $M = S_{-1} \oplus S_i \oplus S_{-i}$.

(c) From Schur's lemma and its corollaries, we have that

$$\dim \operatorname{Hom}_{A\text{-Mod}}(S_i, S_j) = 0$$

if $i \neq j$ and

$$\dim \operatorname{Hom}_{A\text{-Mod}}(S_i, S_i) = 1$$

where $i, j \in \{-1, i, -i\}$.

Let $v_i \in S_i$ be an eigenvector so that $\mathcal{B}_i = \{v_i\}$ is a basis for S_i, for each $i \in \{-1, i, -i\}$; notice that $\mathcal{B}' = \bigcup_i \mathcal{B}_i$ is a basis (of eigenvectors) for M. Then for each i, $\dim \operatorname{Hom}_{A\text{-Mod}}(S_i, M) = 1$, spanned by

$$\alpha_i \colon S_i \to M = \bigoplus_j S_j$$

with $\alpha_i(v_i) = v_i$.

We can extend α_i to an endomorphism of M by defining $\hat{\alpha}_i \colon M \to M$,

$$\hat{\alpha}_i(v_j) = \delta_{ij}\alpha_i(v_i) = \delta_{ij}v_j.$$

These are clearly linearly independent in $\operatorname{End}_A(M)$. Since

$$\dim \operatorname{End}_A(M) = \dim \operatorname{Hom}_{A\text{-Mod}}(M, M)$$
$$= \sum_i \dim \operatorname{Hom}_{A\text{-Mod}}(S_i, M) = 3$$

the set $\{\hat{\alpha}_i \colon M \to M \mid i \in \{-1, i, -i\}\}$ must be a basis for $\operatorname{End}_A(M)$.

We can subcontract the legwork for the first two parts (but not the third!) to SageMath, of course.

```
1   R = MatrixSpace(CC,3)
2   X = R.matrix([[-1,0,1-I],[1+I,I,-1+3*I],[0,0,-I]])
3   X^4
4   X.characteristic_polynomial()
5   X.eigenvalues()
```

4.4 Indecomposable modules

Next, we consider another approach to "building blocks".

Definition 4.4.1. An A-module M is *indecomposable* if it cannot be written as a direct sum of two non-zero submodules. Otherwise M is called *decomposable*.

That is, we look at M and ask if we can decompose it as a direct sum in a non-trivial way ($0 \oplus M$ and $M \oplus 0$ don't count). If not, M is indecomposable. If so, do it and continue in the same way for the submodules in this decomposition. Then every module is a direct sum of indecomposable submodules—right??

Wrong! Much more care is needed: why should this process terminate? Indeed, for an arbitrary module, it need not. Fortunately, we do have:

Proposition 4.4.2. *Let M be a non-zero A-module of finite length. Then M has a decomposition as a finite direct sum of indecomposable submodules.*

Proof. If M is a module of length 1, then M is simple. Its only submodules are 0 and itself, so it cannot be written as a direct sum of two non-zero submodules. So consider the case where M has length at least 2 and for an inductive hypothesis, assume that every module of length strictly less than the length of M can be written as a finite direct sum of indecomposable submodules.

Now either M is indecomposable, in which case we are done, or there exist non-zero submodules N and P such that $M = N \oplus P$. As proper submodules have strictly smaller length (Proposition 4.3.7), by the inductive hypothesis both N and P have decompositions as a finite direct sum of indecomposable submodules and hence M does. $\square$

Remark 4.4.3. Every simple module is indecomposable, as in the first line of this proof. However, not every indecomposable module is simple. Concrete examples will be given in later chapters but the point is that for "most" algebras, there exist non-split extensions that yield indecomposable modules. Indeed, the situation when the decomposition into indecomposables is a decomposition into simples is exactly when the module is semisimple—and not all modules are semisimple.

We see that the definition of indecomposability does not easily lend itself to checking whether or not a given module is indecomposable; *a priori* one has to examine all of its submodules and see whether direct complements exist. However, as with Schur's lemma, indecomposability is detected by the endomorphism algebra: the following results, which we give without proof (referring the reader to [EH, 7.2]), are known as Fitting's lemmata.[7]

Lemma 4.4.4 (Fitting's lemmata). *Let A be an R-algebra over a commutative unital ring R and let M be an A-module of finite length.*

(i) *The module M is indecomposable if and only if every endomorphism $\theta \in \mathrm{End}_A(M)$ is either an isomorphism or is nilpotent (i.e. there exists n such that $\theta^n = 0$).*

(ii) *The module M is indecomposable if and only if the set of endomorphisms of M that do not have a left inverse[8] is a left ideal of $\mathrm{End}_A(M)$.*

(iii) *The module M is indecomposable if and only if the endomorphism algebra $\mathrm{End}_A(M)$ is local, i.e. has a unique maximal ideal I.*

This leads us to one of the cornerstone theorems[9] of representation theory:

Theorem 4.4.5 (Krull–Remak–Schmidt theorem). *Let M be a non-zero A-module of finite length and let*

$$M = M_1 \oplus M_2 \oplus \cdots \oplus M_r = N_1 \oplus N_2 \oplus \cdots \oplus N_s$$

be two decompositions of M into indecomposable submodules. Then $r = s$ and there exists $\sigma \in S_r$ such that $M_i \cong N_{\sigma(i)}$ for all i.

Proof. We follow the proof given in [EH, 7.3]; details omitted here may be found there.

The strategy is as follows. We work by induction on r and note that the base case $r = 1$ is when M is indecomposable and so we can only have

---

[7]"Lemmata" is the plural of "lemma". It would be more accurate to say that different people refer to one or other of the equivalent conditions of Lemma 4.4.4 as Fitting's lemma. To acknowledge the varying usage—and because the word "lemmata" isn't used often enough—we have chosen the slightly unconventional name.

[8]This ideal is called the *Jacobson radical*; we will not discuss it further in full generality but see Proposition 5.3.24 for an alternative description of the Jacobson radial for path algebras of acyclic quivers.

[9]See [Sha] for an explanation of the somewhat complicated history of the name of this result.

$M_1 = M = N_1$. Then for $r > 1$, we first show that there is an i such that $N_1 \cong M_i$. Relabelling if necessary so that $N_1 \cong M_1$, we then show that there is an endomorphism of M sending N_1 to M_1 and N_j to N_j for $j \neq 1$. Taking the quotient by M_1 will then allow us to reduce to modules with fewer summands and hence apply the inductive hypothesis.

For the first step, define the following maps:

- let $\mu_i \colon M \to M_i$ be the projection onto M_i;

- let $\iota_i \colon M_i \to M$ be the inclusion of M_i into M;

- let $\nu_i \colon M \to N_i$ be the projection onto N_i; and

- let $\kappa_i \colon N_i \to M$ be the inclusion of N_i into M.

From the earlier discussion about direct sums and split short exact sequences, we see that $\mu_i \circ \iota_i = \mathrm{id}_{M_i}$ and $\nu_i \circ \kappa_i = \mathrm{id}_{N_i}$. Defining $e_i = \iota_i \circ \mu_i$, we have that e_i is an endomorphism of M with image M_i and kernel $\bigoplus_{j \neq i} M_j$. Then the e_i are orthgonal idempotents[10] such that $\sum_i e_i = \mathrm{id}_M$. Similarly, defining $f_i = \kappa_i \circ \nu_i$ gives another decomposition of the identity map into a sum of orthogonal idempotents, as $\mathrm{id}_M = \sum_i f_i$.

Now $\mathrm{id}_{N_1} = \sum_i \nu_1 \circ e_i \circ \kappa_1$ and Fitting's lemmata imply that one of these summands, $\nu_1 \circ e_i \circ \kappa_1$ say, is an isomorphism of N_1 with M_i. Relabel if necessary so that $i = 1$.

One may show that defining $\gamma \colon M \to M$ by $\gamma = \mathrm{id}_M - f_1 + e_1 \circ f_1$ yields an isomorphism of M with itself such that $\gamma(N_1) = M_1$ and $\gamma(N_j) = N_j$ for $j \neq 1$.

Now since isomorphisms take direct sum decompositions to direct sum decompositions, from γ we obtain

$$M = \gamma(M) = \gamma(N_1) \oplus \gamma(N_2) \oplus \cdots \oplus \gamma(N_s) = M_1 \oplus N_2 \oplus \cdots \oplus N_s$$

By the first isomorphism theorem,

$$M_2 \oplus \cdots \oplus M_r \cong M/M_1 \cong N_2 \oplus \cdots \oplus N_s$$

Let ψ be the composition of these isomorphisms. Then $M_2 \oplus \cdots \oplus M_r$ and

$$\psi^{-1}(N_2 \oplus \cdots \oplus N_s) = \psi^{-1}(N_2) \oplus \cdots \oplus \psi^{-1}(N_s)$$

are decompositions of the same module, to which we may apply the inductive hypothesis and hence conclude the result. $\qquad\square$

This should remind you of the Jordan–Hölder theorem! However—in general (but hold the thought!)—these are not the same result in two guises.

---

[10] We have $e_i^2 = e_i$ (idempotency) and $e_i \circ e_j = 0$ for $i \neq j$ (orthogonality).

4.5   Projective and injective modules

Before we start focusing in on more particular classes of examples, we will look at three final general notions that are important in general representation theory. First is the definition of a free module.

Recall from Section 2.4 that there is a very general categorical notion of a free object in a category, which necessarily applies to A-Mod. Free modules exist for all sets, as is demonstrated by checking that the construction below provides a model, i.e. one can show that there is a left adjoint to the forgetful functor $\mathcal{F}\colon A\text{-Mod} \to \mathbf{Set}$ from the category of A-modules to the category of sets.

Recall from Example 4.1.12 that $_AA$ denotes the regular A-module, i.e. the R-module A with the A-action being left multiplication ($a \triangleright b = ab$).

Then, given a set S, form the A-module $_AA^{\oplus S}$, the S-indexed direct sum of $_AA$ with itself (cf. Definition 4.2.9). One can prove that this is an A-module and that given any function $S \to M$ for $M \in A$-Mod, there exists an A-module homomorphism $_AA^{\oplus S} \to M$, so that $_AA^{\oplus S}$ is the free module on S.

We will mostly be more interested in the following special case:

Definition 4.5.1. Let M be a left A-module. We say M is a *finitely generated free module* if $M \cong {}_AA^{\oplus n}$ for some $n \in \mathbb{N}$.

We say a module M is *finitely generated* if it is the homomorphic image of a finitely generated free module, i.e. there exists $\pi\colon {}_AA^{\oplus n} \twoheadrightarrow M$ for some n. Note that *every* module is the homomorphic image of a free module $_AA^{\oplus S}$ with S not necessarily finite; take $S = M$ with the surjection sending an element $(a_m)_{m \in M}$ of $_AA^{\oplus M}$ to $\sum_m a_m \triangleright m$ in M, for example. So the operative part of the definition is the finiteness.

To make the next definitions, we need to (re-)introduce[11] the Hom-functors. For A an R-algebra and A-modules M and N, the set $\mathrm{Hom}_{A\text{-Mod}}(M, N)$ of module homomorphisms from M to N is an R-module in a natural way.

Firstly, as described in more detail in Section 4.2.1, $\mathrm{Hom}_{A\text{-Mod}}(M, N)$ is an Abelian group under the operation of pointwise addition of maps (i.e. by defining $(f + g)(m) \overset{\text{def}}{=} f(m) + g(m)$). It is an R-module by setting $(r \triangleright f)(m) = f(r \triangleright m)$ for all $f\colon M \to N$ and $m \in M$. (This was part of Exercise 23, which we said at the time was worth doing.)

Problem 26. Show that for an A-module M, we have

$$\mathrm{Hom}_{A\text{-Mod}}({}_AA, M) \cong M$$

as R-modules. In particular,

$$\mathrm{Hom}_{A\text{-Mod}}({}_AA, {}_AA) \cong {}_AA.$$

---

[11] A more general version appeared in Section 2.5.

Show that the latter isomorphism can be extended to an isomorphism of R-algebras between A and $\mathrm{End}_A(A)^{\mathrm{op}}$, where for an algebra $A = (A, m, u)$, $A^{\mathrm{op}} = (A, m \circ \tau, u)$ for $\tau\colon A \otimes A \to A \otimes A$, $\tau(a \otimes b) = b \otimes a$ (see also Section 2.7).

Challenge 27 (Hard!). Work out how the previous exercise is an instance of the Yoneda lemma (2.5.1) for pre-additive categories (4.2.1).

One may then show that for a module $M \in A\text{-Mod}$ we have a (covariant) functor $\mathrm{Hom}_{A\text{-Mod}}(M, -)\colon A\text{-Mod} \to R\text{-Mod}$ and a (contravariant) functor $\mathrm{Hom}_{A\text{-Mod}}(-, M)\colon A\text{-Mod} \to R\text{-Mod}$. The functor $\mathrm{Hom}_{A\text{-Mod}}(M, -)$ sends an A-module N to the R-module $\mathrm{Hom}_{A\text{-Mod}}(M, N)$ and a morphism $f\colon N \to P$ is sent to the function $f\circ-\colon \mathrm{Hom}_{A\text{-Mod}}(M, N) \to \mathrm{Hom}_{A\text{-Mod}}(M, P)$, $(f \circ -)(g) = f \circ g$. The contravariant version is similar.

The following properties identify when one or other of these Hom-functors associated to a module behaves nicely with respect to short exact sequences.

Definition 4.5.2. Let P be a left A-module. We say P is *projective* if the functor $\mathrm{Hom}_{A\text{-Mod}}(P, -)\colon A\text{-Mod} \to R\text{-Mod}$ is *exact*, i.e. for all short exact sequences $0 \to X \to Y \to Z \to 0$ in A-Mod,

$$0 \to \mathrm{Hom}_{A\text{-Mod}}(P, X) \to \mathrm{Hom}_{A\text{-Mod}}(P, Y) \to \mathrm{Hom}_{A\text{-Mod}}(P, Z) \to 0$$

is a short exact sequence in R-Mod.

Dually,

Definition 4.5.3. Let I be a left A-module. We say I is *injective* if the functor $\mathrm{Hom}_{A\text{-Mod}}(-, I)\colon A\text{-Mod} \to R\text{-Mod}$ is exact, i.e. for all short exact sequences $0 \to X \to Y \to Z \to 0$ in A-Mod,

$$0 \to \mathrm{Hom}_{A\text{-Mod}}(Z, I) \to \mathrm{Hom}_{A\text{-Mod}}(Y, I) \to \mathrm{Hom}_{A\text{-Mod}}(X, I) \to 0$$

is a short exact sequence in R-Mod.

Proposition 4.5.4. *The following are equivalent:*

(i) *P is projective;*

(ii) *given an epimorphism $f\colon M \twoheadrightarrow N$ and morphism $p\colon P \to N$, there exists $\bar{p}\colon P \to M$ such that $p = f \circ \bar{p}$;*

$$
\begin{array}{ccc}
 & & P \\
 & \exists\bar{p}\nearrow & \downarrow p \\
M & \twoheadrightarrow & N \\
 & f &
\end{array}
$$

(iii) *every epimorphism $f\colon M \twoheadrightarrow P$ splits, i.e. there exists $s\colon P \to M$ such that $f \circ s = \mathrm{id}_P$ (s is a section of f);*

(iv) *P is isomorphic to a direct summand of a free module: there exists P', Q A-modules and a set I such that $P' \oplus Q = {}_A A^{\oplus I}$ and $P \cong P'$.*

It follows immediately from this that free modules are projective and that direct sums of projective modules are projective.

Proposition 4.5.5. *The following are equivalent:*

(i) *I is injective;*

(ii) *given a monomorphism $g \colon N \hookrightarrow M$ and morphism $i \colon N \to I$, there exists $\bar{i} \colon M \to I$ such that $i = \bar{i} \circ g$;*

$$
\begin{array}{ccc}
 & & I \\
 & \nearrow & \uparrow \\
\exists\bar{i} & & \Big\uparrow i \\
 & & \\
M & \xleftarrow{\;\;g\;\;} & N
\end{array}
$$

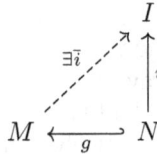

(iii) *every monomorphism $f \colon I \hookrightarrow M$ splits, i.e. there exists $r \colon M \to I$ such that $r \circ f = \mathrm{id}_I$ (r is a retraction of f);*

(iv) *I has a direct complement whenever it exists as a submodule: if $I \leq M$, there exists $Q \leq M$ such that $M = I \oplus Q$.*

Note that it is not true in general that free modules are injective, but this can happen in certain nice situations, one of these being described immediately below and others discussed later.

Remark 4.5.6. The alert reader will observe that the condition 4.5.5(iv) is not the dual of 4.5.4(iv), unlike the other corresponding parts of these Propositions. We have chosen not to state the dual of 4.5.5(iv) and just note here that this is equivalent to the other conditions for projectivity, leaving it as an (advanced) exercise to write down the condition formally and check the claim.

These classes of modules are particularly important when studying finite-dimensional algebras over fields, but the following holds much more generally.

Proposition 4.5.7. *The following are equivalent:*

(i) *the R-algebra A is semisimple;*

(ii) *every module in A-Mod is injective;*

(iii) *every module in A-Mod is projective;*

(iv) *the category A-Mod is semisimple.*

Proof. We first prove (i) if and only if (iv), then (ii) if and only if (iii) and finally (i) if and only if (iii).

- (i)$\Longleftrightarrow$(iv)

  Assume (i). Then every free module $_AA^{\oplus I}$ is also semisimple. Since every module $M \in A$-Mod is a quotient of a free module (e.g. we may take I to be an indexing set for a basis of M) and quotients of semisimple modules are semisimple, we have (iv).

  Conversely (iv) implies (i), by definition.

- (ii)$\Longleftrightarrow$(iii)

  Let M, P and I be A-modules.

  Assume (ii) and let $f: M \twoheadrightarrow P$ be a surjection. Then $\mathrm{Ker}\, f \leq M$ is injective so $M = \mathrm{Ker}\, f \oplus Q$ for some Q. But by the First Isomorphism Theorem, $\mathrm{Im}\, f = P \cong Q$. Then this isomorphism splits f and P is projective.

  Assume (iii) and let $f: I \hookrightarrow M$ be an injection. Then $\mathrm{Coker}\, f = M/I$ is projective so $\pi: M \to M/I$ splits, via $r: M/I \to M$, and $M = \mathrm{Ker}\, \pi \oplus \mathrm{Im}\, r$. But $\mathrm{Ker}\, \pi = I$ so I has a complement in M and is injective.

- (i)$\Longleftrightarrow$(iii)

  Assume (i). Then as above, any $M \in A$-Mod is a quotient of a free module $_AA^{\oplus I}$ via $\pi: {}_AA^{\oplus I} \twoheadrightarrow M$ and this free module is semisimple. Hence $_AA^{\oplus I} = \mathrm{Ker}\, \pi \oplus Q$ with $Q \cong M$, so M is projective.

  Assume (iii). Let M be a submodule of $_AA$. Then since $_AA/M$ is projective, $\pi: {}_AA \to {}_AA/M$ splits and M has a direct complement. By Lemma 4.3.5, $_AA$ is semisimple. $\qquad\square$

Proposition 4.5.8. *Let A be a semisimple R-algebra. Then*

(i) *Every simple module for A is a direct summand of $_AA$.*

(ii) $_AA \cong \bigoplus_{i=1}^n S_i$ *is a finite direct sum of simple modules.*

Proof.

(i) Let $S \in A$-Mod be a simple A-module and let $s \in S \setminus \{0\}$. Then there is an A-module homomorphism $f: {}_AA \to S$ given by $f(a) = a \rhd_S s$ for all $a \in A$; f is linear by the properties of actions and $a \rhd_S (f(b)) = a \rhd_S (b \rhd_S s) = (ab) \rhd_S s = f(ab)$. Now since S is simple and $f(1) = s \neq 0$, $\mathrm{Im}\, f = S$ and f is surjective.

  Then since A semisimple implies S is projective, f splits, i.e. S is (isomorphic to) a direct summand of $_AA$.

(ii) We have that $_AA = \bigoplus_{i \in I} S_i$. Now $1_A \in {}_AA$ so there exists a *finite* subset $J \subseteq I$ and elements $s_j \in S_j$ for all $j \in J$ such that $1_A = \sum_{j \in J} s_j$.

  But then for all $a \in A$, $a = a1_A = a\left(\sum_{j \in J} s_j\right) \in \bigoplus_{j \in J} S_j$. So $\bigoplus_{j \in J} S_j \subseteq \bigoplus_{i \in I} S_i = {}_AA \subseteq \bigoplus_{j \in J} S_j$ and hence $_AA = \bigoplus_{j \in J} S_j$. $\qquad\square$

Note that if A is semisimple, the simple modules generate all of A-Mod by taking direct sums, there are no non-trivial extensions between them and we have Schur's lemma to tell us that any non-zero homomorphism between two simples is an isomorphism. Thus, in a semisimple category, we know all the representation-theoretic information as soon as we can describe the simple modules.

If $\mathbb{K}$ is a field and A is a finite-dimensional semisimple $\mathbb{K}$-algebra, it follows from the above that every simple module for A is finite-dimensional and there are finitely many of these. However, the Proposition holds in much greater generality: even if some or all of the simples are infinite-dimensional (and so A is), the second part tells us there are still finitely many simples (up to isomorphism).

For completeness, we include the following theorem, which is another of the fundamental theorems of representation theory.

Theorem 4.5.9 (Artin–Wedderburn). *If A is a semisimple R-algebra then*

$$A \cong \prod_{i=1}^{r} \mathrm{M}_{n_i}(D_i),$$

a product of matrix rings of dimension n_i over division R-algebras D_i.

Proof. Let $A = \bigoplus_{i=1}^{r} S_i^{\oplus n_i}$ be a decomposition of A into a direct sum of simple modules such that $S_i \not\cong S_j$ for $i \neq j$. Define $B_i = S_i^{\oplus n_i}$ for each i, so $A = \bigoplus_{i=1}^{r} B_i$.

Then since $\mathrm{Hom}_{A\text{-Mod}}(S_i, S_j) = 0$ for all $i \neq j$, it follows that we also have $\mathrm{Hom}_{A\text{-Mod}}(B_i, B_j) = 0$. One may then show (see e.g. [DK, §1.7]) that via Exercise 26, $A \cong \mathrm{End}_A({}_A A)^{\mathrm{op}} \cong \prod_{i=1}^{r} \mathrm{End}_A(B_i)$.

Now defining $D_i = \mathrm{End}_A(S_i)$ (which is a division R-algebra by Schur's lemma, 4.3.10) and since $B_i = S_i^{\oplus n_i}$, $\mathrm{End}_A(B_i) \cong \mathrm{M}_{n_i}(D_i)$ and we have the result. $\qquad\square$

Corollary 4.5.10. *If $\mathbb{K}$ is algebraically closed and A is a semisimple $\mathbb{K}$-algebra then*

$$A \cong \prod_{i=1}^{r} \mathrm{M}_{n_i}(\mathbb{K}).$$

Proof. By Lemma 4.3.11, since S_i is simple and $\mathbb{K}$ algebraically closed, $D_i = \mathrm{End}_A(S_i) \cong \mathbb{K}$. $\qquad\square$

Recall from Example 4.3.3 that $\mathrm{M}_{n_i}(D_i)$ has exactly one simple module (up to isomorphism), namely $D_i^{n_i}$.

4.E   Exercises

Exercise 4.1. Check that the diagrams in the alternative definition of an A-module at the start of Section 4.1 do indeed encode the equations claimed for them immediately afterwards.

Exercise 4.2 (Problem 20). Check that the definition in Example 4.1.9 of the trivial module for $R[x]$ is indeed an $R[x]$-module structure.

Exercise 4.3 (Problem 21). Give a definition of a right module for an R-algebra A and of the right regular module A_A.

Exercise 4.4. Let $\mathbb{K}$ be a field and let A be a $\mathbb{K}$-algebra. Let M be an A-module and $\lambda \in \mathbb{K}$. Define $\varphi_\lambda : M \to M$ by $\varphi_\lambda(m) = \lambda m$ for all $m \in M$.

(a) Prove that φ_λ is an A-module homomorphism.

(b) Prove that φ_λ is an A-module automorphism if and only if $\lambda \neq 0$.

(c) Prove that $\varphi_\lambda \circ \psi = \psi \circ \varphi_\lambda$ for any A-module homomorphism $\psi : M \to M$.

(d) Assume M is finite dimensional. Give the matrix form of φ_λ as a linear transformation from M to M.

Exercise 4.5 (Proof of Proposition 4.2.4). Let A be a $\mathbb{K}$-algebra and let $f : M \to N$ be an A-module homomorphism. Prove that $\operatorname{Ker} f$ and $\operatorname{Im} f$ are submodules of M and N respectively.
Show that the homomorphism f is injective if and only if $\operatorname{Ker} f = 0$, and that f is surjective if and only if $\operatorname{Im} f = N$.

Exercise 4.6. Let A be a $\mathbb{K}$-algebra and let $\varphi : M \to M$ be an A-module endomorphism of an A-module M. Suppose that $\varphi^2 = \varphi$, where $\varphi^2 = \varphi \circ \varphi$ is the composition of φ with itself.
Prove that there exist A-submodules U, V of M such that $M = U \oplus V$ with $\varphi(u) = 0$ for each $u \in U$ and $\varphi(v) = v$ for each $v \in V$.

Exercise 4.7. Prove Proposition 4.2.10.

Exercise 4.8 (Problem 22). Find out about Emmy Noether and tell someone about her.

Exercise 4.9 (Problem 23). Let U, V, W be A-modules for an R-algebra A, let $\varphi_1, \varphi_2 \in \operatorname{Hom}_{A\text{-Mod}}(U, V)$ and let $\psi \in \operatorname{Hom}_{A\text{-Mod}}(V, W)$.

(a) Prove that the composition $\psi \circ \varphi_1 : U \to W$ is an A-module homomorphism.

(b) Prove that the function $(\varphi_1 + \varphi_2) : U \to V$ defined by

$$(\varphi_1 + \varphi_2)(u) = \varphi_1(u) + \varphi_2(u)$$

for all $u \in U$ makes $\operatorname{Hom}_{A\text{-Mod}}(U, V)$ into an Abelian group.

(c) Defining $r \triangleright \varphi_1 \colon U \to V$, $(r \triangleright \varphi_1)(u) = r \triangleright \varphi_1(u)$ for $r \in R$, show that $\mathrm{Hom}_{A\text{-Mod}}(U, V)$ is an R-module.

(d) Suppose that $U = V$. Prove that $\mathrm{End}_A(U) \overset{\mathrm{def}}{=} \mathrm{Hom}_{A\text{-Mod}}(U, U)$ is an algebra for the addition and R-module structure defined above and with the multiplication being the composition of A-homomorphisms. Give the multiplicative and the additive identity elements.

(e) Is $\mathrm{End}_A(U)$ commutative as a ring?

Exercise 4.10. Let A be a $\mathbb{K}$-algebra for which there exists an A-module M such that $M \cong S_1 \oplus S_2 \oplus S_3$ with each S_i simple.

Give a composition series for M.

Exercise 4.11. Let $A = \mathbb{C}[x]$ and let $n \geq 2$. Let I be the ideal $(x^n - 1)$ generated by $x^n - 1$. Let M be the A-module A/I with the natural action of A on the quotient A/I, $a \triangleright (b + I) = ab + I$.

Explicitly, $M = A/I$ is the vector space of dimension n with natural basis $\mathcal{B} = \{m_1, m_2, \ldots, m_n\}$ for $m_i \overset{\mathrm{def}}{=} x^{i-1} + I$ $(1 \leq i \leq n)$ and with action defined by

$$x \triangleright m_i = m_{i+1}$$

where we interpret the indices modulo n.

Using Schur's lemma and subsequent results, show that M is not simple.

Exercise 4.12 (Problem 26). Show that for an A-module M, we have

$$\mathrm{Hom}_{A\text{-Mod}}(M, {}_A A) \cong M$$

as R-modules. In particular,

$$\mathrm{Hom}_{A\text{-Mod}}({}_A A, {}_A A) \cong {}_A A$$

Show that the latter isomorphism can be extended to an isomorphism of R-algebras between A and $\mathrm{End}_A(A)^{\mathrm{op}}$, where for an algebra $A = (A, m, u)$, $A^{\mathrm{op}} = (A, m \circ \tau, u)$ for $\tau \colon A \otimes A \to A \otimes A$, $\tau(a \otimes b) = b \otimes a$ (see also Section 2.7).

Exercise 4.13 (Challenge 27; hard!). Work out how Exercise 4.12 is an instance of the Yoneda lemma (2.5.1) for pre-additive categories (4.2.1).

Exercise 4.14. Prove Proposition 4.5.4 and Proposition 4.5.5.

Chapter 5

Examples

In this chapter, we will use the general results of the previous one in various different situations. We start with the case of principal ideal domains and use the ideas we have seen to give a classification of finitely generated modules over such rings. As a special case, this yields the classification of finitely generated Abelian groups.

Next, we turn back to groups. We introduce an algebra, the group algebra, whose representation theory is the same (in a strong formal sense) as that of the group. Then we can identify when this algebra is semisimple, which turns out to be most of the time, so that the previous results tell us about the representation theory in that case.

Thirdly, we study the representation theory of quivers in terms of modules in a similar way. Again, we can introduce an algebra, the path algebra, whose representation theory is the same as that of the quiver. In contrast, semisimplicity is rare in this case but we will see that there are nevertheless things we can say.

5.1   Modules for principal ideal domains

Recall from Section 1.3.9 that an ideal I in a commutative ring R is principal if $I = aR$ for some $a \in R$. A *principal ideal domain* (or PID for short) is a ring R that is an integral domain and in which every ideal is principal; more background and detail than in Section 1.3.9 can be found in [AF], for example.

Examples include fields, the ring of integers $\mathbb{Z}$ and polynomial rings $\mathbb{K}[x]$ over a field.

Principal ideal domains are in particular *unique factorization domains* (abbreviated to UFDs). Every Euclidean domain is a PID (but not vice versa); being a Euclidean domain means that one can always calculate greatest common divisors, whereas in a PID one only knows that these exist.

From the point of view of this book, the most important reason for focussing on PIDs is that it is possible to give a classification of the finitely generated modules over a PID. This result has a number of important applications for both the structure theory of groups (since $\mathbf{Ab} = \mathbb{Z}\text{-Mod}$) and linear algebra, in the shape of "normal forms" for matrices.

For the remainder of this section, assume that R is a principal ideal domain.

In this section, we apply the results of Chapter 4 in the case $A = R$, that is, considering R as an R-algebra as seen before, with the R-module structure being that of $_R R$, the regular module, and with the multiplication and unit of R defining the algebra structure.

Recall from Section 4.5 that an R-module M is free if $M \cong {}_R R^{\oplus I}$ for some set I, a finitely generated free module if $M \cong {}_R R^{\oplus n}$ for some n and finitely generated if it is a quotient of a finitely generated free module.

Definition 5.1.1. Let $M \in R$-Mod. We say M is *cyclic* if it is generated by one element, i.e. there exists $m \in M$ such that $M = Rm = \{r \triangleright m \mid r \in R\}$.

Although "1-generator" might be a more descriptive name (some texts even say "monogeneous") we use "cyclic" as it is meant to remind us of cyclic groups and cyclic subgroups. Indeed, as we will see, these are special cases.

Proposition 5.1.2. *Let R be a PID and M an R-module. Then M is cyclic if and only if $M \cong R/Rd$ for some $d \in R$.*

Proof. If M is cyclic, there exists $m \in M$ such that $M = Rm$ and hence there is a surjection $\pi \colon {}_R R \twoheadrightarrow M$ given by $r \mapsto rm$. However all we need to observe is that π has a kernel which is an ideal of R and $R/\operatorname{Ker} \pi \cong M$. Since R is a PID, the kernel is a principal ideal: $\operatorname{Ker} \pi = Rd$ for some $d \in R$ and hence $R/Rd \cong M$.

For the converse, the coset $1 + Rd$ generates R/Rd since for any coset $r + Rd \in R/Rd$ we have $r \triangleright (1 + Rd) = r + Rd$. Hence R/Rd is cyclic. $\qquad\square$

Observe that if M is cyclic, so $M \cong R/Rd$ for some $d \in R$, different things may happen depending on d. If $d = 0$, $Rd = 0$ and $M \cong R/0 = R$. Conversely if d is a unit (i.e. invertible in R), then $Rd = R$ since $1 = d^{-1}d \in Rd$ and so $M \cong R/R = 0$.

We can also say when two cyclic modules are isomorphic:

Lemma 5.1.3. *Let R be a PID and M, N cyclic R-modules. Let $d, d' \in R$ be such that $M \cong R/Rd$ and $N \cong R/Rd'$. Then $M \cong N$ if and only if d and d' are associates, i.e. there exists a unit $u \in R^\times$ such that $d' = ud$.*

Proof. If $M \cong N$, let $\varphi \colon R/Rd \to R/Rd'$ be the corresponding isomorphism of R-modules. Then there exists $s \in R$ such that $\varphi(1 + Rd) = s + Rd'$. Now

$$
\begin{aligned}
\varphi(d' + Rd) &= \varphi(d' \triangleright (1 + Rd)) \\
&= d' \triangleright \varphi(1 + Rd) \\
&= d's + Rd' \\
&= Rd' \\
&= 0 + Rd' \\
&= \varphi(0 + Rd)
\end{aligned}
$$

Since φ is injective, we deduce that $d' + Rd = 0 + Rd$ so $d' \in Rd$. Similarly, using φ^{-1} we deduce that $d \in Rd'$.

Hence there exist $a, b \in R$ such that $d' = ad$ and $d = bd'$. Then $d' = abd'$ and R being a domain implies $ab = 1$. So a is a unit such that $d' = ad$ and d, d' are associates.

Conversely, if d, d' are associates, so there exists $u \in R^\times$ such that $d' = ud$, then for any $rd' \in Rd'$, $rd' = rud \in Rd$ and for any $rd \in Rd$, $rd = ru^{-1}d' \in Rd'$. Hence $Rd = Rd'$ and so $M \cong R/Rd = R/Rd' \cong N$. $\qquad\square$

Example 5.1.4. Let $R = \mathbb{Z}$. Recall that $\mathbb{Z}$-Mod $=$ **Ab**, so that when we are talking about $\mathbb{Z}$-modules we are talking about Abelian groups at the same time. Then the above results say that every cyclic $\mathbb{Z}$-module (Abelian group) is isomorphic to $\mathbb{Z}/d\mathbb{Z}$ for some $d \in \mathbb{Z}$ and furthermore $\mathbb{Z}/d\mathbb{Z} \cong \mathbb{Z}/d'\mathbb{Z}$ if and only if $d = \pm d'$, since $\mathbb{Z}^\times = \{\pm 1\}$.

We usually write $\mathbb{Z}_d = \mathbb{Z}/d\mathbb{Z}$ for the quotient group, which is exactly the group of integers modulo d. So the cyclic Abelian groups are $\mathbb{Z}/0\mathbb{Z} = \mathbb{Z}, \mathbb{Z}/1\mathbb{Z} = 0, \mathbb{Z}_2, \mathbb{Z}_3, \ldots$.

Note that we have been using left modules throughout so would write $\mathbb{Z}d$ for consistency with the above, but it is much more common to write $d\mathbb{Z}$ for the set of multiples of d. Since $\mathbb{Z}$ is a commutative ring, no harm is done.

Now comes the main theorem.

Theorem 5.1.5 (cyclic decomposition). *Let R be a PID. Every finitely generated R-module M is isomorphic to a direct product of cyclic R-modules*

$$M \cong R/Rd_1 \times R/Rd_2 \times \cdots \times R/Rd_n$$

where d_1 is not a unit and $d_i | d_{i+1}$ for $1 \leq i \leq n - 1$. Furthermore, the d_i are unique up to associates.

Definition 5.1.6. The elements d_i in a cyclic decomposition of M as in the theorem are called the *elementary divisors* of M.

We prove the cyclic decomposition theorem via a sequence of results as below.

Lemma 5.1.7. *Let $F = {}_R R^{\oplus n}$ be a finitely generated free R-module with basis $\mathcal{B} = \{b_1, \ldots, b_n\}$. Let $d_1, \ldots, d_n \in R$ and set N to be the submodule of F generated by $\{d_1 b_1, \ldots, d_n b_n\}$. Then*

$$F/N \cong R/Rd_1 \times \cdots \times R/Rd_n$$

is a direct product of cyclic R-modules.

Proof. There is a well-defined surjective R-module homomorphism $\psi \colon F \to R/Rd_1 \times \cdots \times R/Rd_n$ given by

$$\psi(r_1 b_1 + r_2 b_2 + \cdots + r_n b_n) = (r_1 + Rd_1, r_2 + Rd_2, \ldots, r_n + Rd_n)$$

whose kernel is exactly N. $\qquad\square$

Proposition 5.1.8. *Let R be a PID. Every submodule of a finitely generated free R-module is finitely generated.*

We will not prove this now, but instead prove a more general statement in Section 6.1. Note that this is false without some assumptions on R (it will turn out that being a PID is more than is needed, but is certainly sufficient).

Now, our finitely generated module M is isomorphic to a quotient F/N where F is a finitely generated free module and (by the proposition) N is a finitely generated submodule of F. We would like to apply Lemma 5.1.7 to see that $M \cong F/N$ is isomorphic to a direct product of cyclic R-modules, then it will remain to show that the d_i may be chosen (uniquely up to associates) as in the statement of Theorem 5.1.5.

This line of argument requires proving that we can choose a pair of bases $\tilde{\mathcal{B}}$ and $\tilde{\mathcal{C}}$ for F and N respectively such that (i) $\tilde{\mathcal{B}} = \{\tilde{b}_1, \ldots, \tilde{b}_n\}$ and $\tilde{\mathcal{C}} = \{\tilde{c}_1 \qquad = \qquad d_1\tilde{b}_1, \ldots, \tilde{c}_n = d_n\tilde{b}_n\}$ for some $d_i \in R$ (so that 5.1.7 applies) and (ii) d_1 is not a unit and $d_i|d_{i+1}$.

From the natural basis $\mathcal{B} = \{b_1, \ldots, b_n\}$ of F and a generating set $\mathcal{C} = \{c_1, \ldots, c_n\}$ for N (which exists by 5.1.8) we can write down a matrix with entries from R expressing the c_i in terms of the b_i, just as we would do in linear algebra. That is, let C be the matrix with entries $c_{ij} \in R$ where c_{ij} is defined by $c_i = \sum_{j=1}^{n} c_{ij}b_j$.

Let us say a (possibly non-square) matrix D is *diagonal* if $d_{ij} = 0$ for all $i \neq j$. Our aim is to find a diagonalization of the matrix C such that the resulting diagonal entries satisfy $d_i|d_{i+1}$. For then we will have changed the basis $\mathcal{B}$ and generating set $\mathcal{C}$, via invertible matrices over R, into a basis $\tilde{\mathcal{B}}$ and generating set $\tilde{\mathcal{C}}$ with the desired property. (One can check that the basic propositions of linear algebra—that is, the theory of $\mathbb{K}$-modules—extend to R-modules so that multiplying by an invertible matrix over R preserves the properties of being a generating set or being a basis.)

That is, the cyclic decomposition theorem will follow if we can prove the next theorem, which claims the existence of the Smith Normal Form.

Theorem 5.1.9 (Smith Normal Form). *Let R be a PID and let C be an $n \times t$ matrix with entries in R. Then there exist invertible matrices P and Q over R such that PCQ is diagonal with diagonal entries d_i ($1 \leq i \leq \min\{n,t\}$) satisfying $d_i|d_{i+1}$.*

Proof. The argument we use follows that set out in [Jac, Section 3.7], where further details can be found. The strategy is algorithmic: given a matrix, in a sequence of steps one can turn it into the desired form, using *elementary matrix operations*.

Abusing notation somewhat, let e_{ij} denote an elementary matrix whose (i,j) entry is 1_R and is zero elsewhere. Below, e_{ij} will be either $n \times n$ or $r \times r$, depending on whether we act with it on the left or right, but we will

not burden the notation further by including its size explicitly. Note that
$I = \sum_i e_{ii}$.

- For $r \in R$ and $i \neq j$, set $T_{ij}(r) = I + re_{ij}$. Left multiplication of
  a matrix C by $T_{ij}(r)$ gives the matrix whose ith row is replaced by
  the ith row plus r times the jth row of C, the rest being unchanged.
  Right multiplication is similar with i and j interchanged and "column"
  instead of "row".

- For $u \in R^\times$, set $D_i(u) = I + (u-1)e_{ii}$. Left multiplication by $D_i(u)$ mul-
  tiplies the ith row by u; likewise for right multiplication and columns.

- Let $P_{ij} = I - e_{ii} - e_{jj} + e_{ij} + e_{ji}$. Left multiplication by P_{ij} interchanges
  the ith and jth rows; right multiplication the corresponding columns.

- Let $A \in M_2(R)^\times$ be an invertible 2×2 matrix over R. Let $D(A) = A \oplus I$.

Each of the above matrices is invertible: $T_{ij}(r)T_{ij}(-r) = I$, $D_i(u)D_i(u^{-1}) = I$, $P_{ij}^2 = I$ and $D(A)D(A^{-1}) = I$. So it suffices to show that by left and right
multiplication by suitably chosen matrices of the form $T_{ij}(r)$, $D_i(u)$, P_{ij} and
$D(A)$, we can manipulate our given matrix C into diagonal form. For then
multiplying together the constituent matrices will give the P and Q we seek.

Since R is a PID, we have unique factorization of elements of R into
primes. Let $\Omega(r)$ denote the number of prime factors of r counted with multiplicity; so if $r = \prod_i p_i^{m_i}$, then $\Omega(r) = \sum_i m_i$. By convention, if u is a unit,
we set $\Omega(u) = 0$.

If $C = 0$ there is nothing to prove. Let c_{ij} be a non-zero element of C
with minimal $\Omega(c_{ij})$. Using elementary transformations of the form P_{ij} we
bring this element to the $(1,1)$ position, so now we may assume that $c_{11} \neq 0$
and $\Omega(c_{11}) \leq \Omega(c_{ij})$ for all $c_{ij} \neq 0$.

Assume that $c_{11} \nmid c_{1k}$ for some k. By swapping the second and kth
columns, we may assume that $c_{11} \nmid c_{12}$. Then the greatest common divisor (which exists since we are in a PID) of c_{11} and c_{12}, $d = (c_{11}, c_{12})$
has $\Omega(d) < \Omega(c_{11})$. By Bézout's identity, there exist $x, y \in R$ such that
$c_{11}x + c_{12}y = d$. Since $d | c_{11}, c_{12}$ there exist $s, t \in R$ such that $c_{11} = -td$
and $c_{12} = sd$.

Now

$$\begin{pmatrix} -t & s \\ y & -x \end{pmatrix} \begin{pmatrix} x & s \\ y & t \end{pmatrix} = \begin{pmatrix} sy - tx & 0 \\ 0 & sy - tx \end{pmatrix} = \begin{pmatrix} 1 & 0 \\ 0 & 1 \end{pmatrix}$$

since $d(sy - tx) = sdy - tdx = c_{12}y + c_{11}x = d$ and R is an integral domain.
Hence the matrix $A = \begin{pmatrix} x & s \\ y & t \end{pmatrix}$ is invertible.

Then multiplying C on the right by $D(A)$ transforms C to have first row
$(d, 0, c_{13}, \ldots, c_{1t})$ with $\Omega(d) < \Omega(c_{11})$. Repeating this process as necessary,
we can assume that in the first row of C, $c_{11} \neq 0$, $\Omega(c_{11}) < \Omega(c_{ij})$ for all
$c_{ij} \neq 0$ and $c_{11} | c_{1k}$ for all k. The same argument but multiplying on the

left at appropriate points means we can extend this to include the condition that $c_{11}|c_{k1}$ for all k. Let us call this suite of manipulations "ensuring divisibility".

From this point, we may use elementary transformations $T_{ij}(r)$ for suitable r to manipulate C into the form

$$\begin{pmatrix} c_{11} & 0 & 0 & \cdots & 0 & 0 \\ 0 & c_{22} & c_{23} & \cdots & c_{2(t-1)} & c_{2t} \\ 0 & c_{32} & c_{33} & \cdots & c_{3(t-1)} & c_{3t} \\ \vdots & \vdots & \vdots & & \vdots & \vdots \\ 0 & c_{n2} & c_{n3} & \cdots & c_{n(t-1)} & c_{nt} \end{pmatrix}$$

(Since $c_{11}|c_{1k}$, so there exists r such that $c_{1k} = rc_{11}$, applying $T_{1k}(-r)$ on the right replaces c_{1k} by $c_{1k} + (-r)c_{11} = 0$.)

If for some $k, l, c_{11} \nmid c_{kl}$ then transforming this matrix by applying $T_{1k}(1)$ on the left, we can run the "ensuring divisibility" process again (and again, and again...) until we obtain a C of the above form with $c_{11}|c_{kl}$ for all k, l.

Repeating all of this process on the submatrix given by deleting the first row and column now yields a matrix of the form

$$\begin{pmatrix} c_{11} & 0 & 0 & \cdots & 0 & 0 \\ 0 & c_{22} & 0 & \cdots & 0 & 0 \\ 0 & 0 & c_{33} & \cdots & c_{3(t-1)} & c_{3t} \\ \vdots & \vdots & \vdots & & \vdots & \vdots \\ 0 & 0 & c_{n3} & \cdots & c_{n(t-1)} & c_{nt} \end{pmatrix}$$

with $c_{11}|c_{22}|c_{kl}$ for all k, l.

Continuing in this way, we eventually reach a diagonal matrix with the desired divisibility of the diagonal entries. Note that this algorithm *does* terminate in finitely many steps because for any given matrix C, the set $\{\Omega(c_{ij})\}$ is bounded above and the taking of greatest common divisors where needed lowers Ω strictly. $\qquad\square$

SageMath can find the elementary divisors and Smith normal form over a PID: here are three examples.

```
1  C=matrix(ZZ,4,6,range(24));
2  C.elementary_divisors()
3  D,P,Q=C.smith_form(); (D,P,Q)
```

tells us that the matrix

$$C = \begin{pmatrix} 0 & 1 & 2 & 3 & 4 & 5 \\ 6 & 7 & 8 & 9 & 10 & 11 \\ 12 & 13 & 14 & 15 & 16 & 17 \\ 18 & 19 & 20 & 21 & 22 & 23 \end{pmatrix}$$

has elementary divisors 1, 6, 0 and 0 and that its Smith normal form

$$D = \begin{pmatrix} 1 & 0 & 0 & 0 & 0 & 0 \\ 0 & 6 & 0 & 0 & 0 & 0 \\ 0 & 0 & 0 & 0 & 0 & 0 \\ 0 & 0 & 0 & 0 & 0 & 0 \end{pmatrix}$$

is obtained by computing PCQ for

$$P = \begin{pmatrix} 0 & 0 & 3 & -2 \\ 0 & 0 & -1 & 1 \\ 1 & 0 & -3 & 2 \\ 0 & 1 & -2 & 1 \end{pmatrix}$$

and

$$Q = \begin{pmatrix} 0 & 0 & 1 & 0 & 0 & 0 \\ 0 & 0 & 0 & 1 & 0 & 0 \\ 0 & 0 & 0 & 0 & 1 & 0 \\ 0 & 0 & 0 & 0 & 0 & 1 \\ -1 & 5 & -5 & -4 & -3 & -2 \\ 1 & -4 & 4 & 3 & 2 & 1 \end{pmatrix}$$

For an example over $R = \mathbb{Q}[x]$, let us take

$$C = \begin{pmatrix} x+1 & 3 \\ x-2 & x+1 \end{pmatrix}$$

Then SageMath tells us via

```
1  R.<x>=QQ[];
2  C1=x*matrix(R,2,2,[1,0,1,1]);
3  C0=matrix(R,2,2,[1,3,-2,1]);
4  C=C1+C0;
5  C.elementary_divisors()
6  D,P,Q=C.smith_form(); (D,P,Q)
```

that C has elementary divisors 1 and $x^2 - x + 7$.

To identify an example with repeated elementary divisors, a little experimenting (mixed with some reflection on the above proof) enables us to find

$$C = \begin{pmatrix} x+1 & x^2+2x+1 \\ 0 & x+1 \end{pmatrix}$$

whose elementary divisors are $x+1$, $x+1$, via

```
1  R.<x>=QQ[];
2  C2=x*x*matrix(R,2,2,[0,1,0,0]);
3  C1=x*matrix(R,  2,  2,  [1,2,0,1]);
4  C0=matrix(R,  2,  2,  [1,1,0,1]);
5  C=C2+C1+C0;
6  C.elementary_divisors()
```

(We defined C degree by degree with separate variable names Ci to make the code more readable, but we could have just run the sum together on a single line to define C in one go.)

Continuing the example above, we may take $R = \mathbb{Z}$ in the cyclic decomposition theorem to obtain:

Corollary 5.1.10. *Every finitely generated Abelian group is isomorphic to*

$$\mathbb{Z}/d_1\mathbb{Z} \times \mathbb{Z}/d_2\mathbb{Z} \times \cdots \times \mathbb{Z}/d_n\mathbb{Z}$$

where $d_1 \neq \pm 1$ and $d_i | d_{i+1}$ for $1 \leq i \leq n - 1$.

Note that $d|0$ for any d so components isomorphic to $\mathbb{Z}$ are certainly allowed in the case of *finitely generated* Abelian groups. If we force all of the d_i to be strictly positive, we obtain a finite Abelian group; indeed, we obtain them all this way.

Corollary 5.1.11. *Every finite Abelian group is isomorphic to*

$$\mathbb{Z}/d_1\mathbb{Z} \times \mathbb{Z}/d_2\mathbb{Z} \times \cdots \times \mathbb{Z}/d_n\mathbb{Z}$$

where $d_1 \neq \pm 1$, $d_i > 0$ for all i and $d_i | d_{i+1}$ for $1 \leq i \leq n - 1$.

Note that the order of $\prod_{i=1}^{n} \mathbb{Z}/d_i\mathbb{Z}$ is $\prod_{i=1}^{n} d_i$.

At the other extreme, if $d_i = 0$ for all i, we obtain $\mathbb{Z}^n$. These are exactly the finitely generated *free* Abelian groups.

Example 5.1.12. The Abelian groups of order $108 = 2^2 \cdot 3^3$ are determined by tuples $(d_1, \ldots, d_n)$ with $d_i > 0$, $d_1 > 1$, $d_i | d_{i+1}$ and $\prod_i d_i = 108$. These are:

$(3, 3, 12)$: $\mathbb{Z}_3 \times \mathbb{Z}_3 \times \mathbb{Z}_{12}$

$(3, 6, 6)$: $\mathbb{Z}_3 \times \mathbb{Z}_6 \times \mathbb{Z}_6$

$(6, 18)$: $\mathbb{Z}_6 \times \mathbb{Z}_{18}$

$(3, 36)$: $\mathbb{Z}_3 \times \mathbb{Z}_{36}$

$(2, 54)$: $\mathbb{Z}_2 \times \mathbb{Z}_{54}$

(108): $\mathbb{Z}_{108}$

Our second application is to canonical forms in linear algebra. By taking $\mathbb{K}$ a field and $R = \mathbb{K}[x]$, since polynomial rings over fields are PIDs, we can apply the cyclic decomposition theorem to describe every finitely generated $\mathbb{K}[x]$-module as a product of cyclic modules $\prod_i \mathbb{K}[x]/\mathbb{K}[x]d_i$ where the d_i are monic polynomials ($\mathbb{K}[x]^\times = \mathbb{K}^\times = \mathbb{K} \setminus \{0\}$), $d_1 \neq 1$ and $d_i | d_{i+1}$.

However, there is another way to construct *finite-dimensional* $\mathbb{K}[x]$-modules, in terms of linear maps.

Lemma 5.1.13. *Let V be a finite-dimensional vector space over a field $\mathbb{K}$ and let $\alpha: V \to V$ be a linear map.*

Then the action $f \triangleright v = f(\alpha)(v)$ for all $f \in \mathbb{K}[x]$, $v \in V$ makes V into a finite-dimensional $\mathbb{K}[x]$-module, which we will call V_α.

Conversely, every finite-dimensional $\mathbb{K}[x]$-module is isomorphic to V_α for some $\alpha: V \to V$.

Proof. Exercise. For the converse direction, show that if V is a $\mathbb{K}[x]$-module, the map $x \triangleright -: V \to V$, $(x \triangleright -)(v) = x \triangleright v$ is a linear map and that (by inducting on the degree of f) this recovers the action $f \triangleright v = f(\alpha)(v)$. $\square$

The relationship between the cyclic decomposition theorem and linear algebra begins with:

Lemma 5.1.14. *Let V_α be a finite-dimensional $\mathbb{K}[x]$-module, with elementary divisors $(d_1, \ldots, d_n)$. Then d_n is the minimal polynomial of α.*

Now we can explain two canonical matrix forms, *rational canonical form* and *Jordan normal form*.

Rational canonical form

Let V_α be a finite-dimensional $\mathbb{K}[x]$-module and assume V_α is cyclic, so $V_\alpha \cong \mathbb{K}[x]/\mathbb{K}[x]d$ for d the minimal polynomial of α. We may write

$$d = x^n - a_{n-1}x^{n-1} - \cdots - a_1 x - a_0$$

with $a_i \in \mathbb{K}$.

Let $\mathcal{B} = \{x^i + \mathbb{K}[x]d \mid 0 \leq i \leq n-1\}$ be the natural (ordered) basis for V_α and define $C(d)$, the *companion matrix* of d, to be the matrix of α with respect to this basis.

Then

$$C(d) = \begin{pmatrix} 0 & 0 & 0 & \cdots & 0 & a_0 \\ 1 & 0 & 0 & \cdots & 0 & a_1 \\ 0 & 1 & 0 & \cdots & 0 & a_2 \\ \vdots & \vdots & \vdots & & \vdots & \vdots \\ 0 & 0 & 0 & \cdots & 1 & a_{n-1} \end{pmatrix}$$

Now given any $V_\alpha \cong \prod_i \mathbb{K}[x]/\mathbb{K}[x]/d_i$, each $V_{\alpha,i} \stackrel{def}{=} \mathbb{K}[x]/\mathbb{K}[x]d_i$ is a submodule and choosing the basis $\mathcal{B}_i = \{x^j + \mathbb{K}[x]d_i\}$, we obtain a basis for V_α

with respect to which the matrix of α is the direct sum of the companion matrices $C(d_i)$, i.e.

$$
C(\alpha) \stackrel{\text{def}}{=} \begin{pmatrix} C(d_1) & 0 & \cdots & 0 \\ 0 & C(d_2) & \cdots & 0 \\ \vdots & \vdots & & \vdots \\ 0 & 0 & \cdots & C(d_n) \end{pmatrix}
$$

This is the *rational canonical form* of α. Furthermore, linear maps α and β are similar if and only if $C(\alpha) = C(\beta)$.

We also see from the rational canonical form that the characteristic polynomial of α is equal to the product of the elementary divisors of α, $\prod_i d_i$.

SageMath can find the rational canonical form over a field, as follows.

```
1  R = MatrixSpace(QQ,3,3)
2  M = R.matrix([[1,2,3],[4,5,6],[7,8,9]])
3  M.minimal_polynomial().factor()
4  M.characteristic_polynomial().factor()
5  M.rational_form()
```

returns that the minimal polynomial equals the characteristic polynomial, which factors into irreducibles as $x(x^2 - 15x - 18)$, and that the rational canonical form of M is

$$
\begin{pmatrix} 0 & 0 & 0 \\ 1 & 0 & 18 \\ 0 & 1 & 15 \end{pmatrix}
$$

Jordan normal form

Let $\mathbb{K}$ be a field and let $\alpha \colon V \to V$ be a linear map whose eigenvalues all belong to $\mathbb{K}$. (So if $\mathbb{K}$ is algebraically closed, e.g. $\mathbb{K} = \mathbb{C}$, α can be arbitrary.)

Now, using the fact that $\mathbb{K}[x]$ is a PID and hence a UFD, the factorization $d = p_1^{m_1} p_2^{m_2} \cdots p_r^{m_r}$ imples that the cyclic $\mathbb{K}[x]$-module $\mathbb{K}[x]/\mathbb{K}[x]d$ can be decomposed as

$$
\mathbb{K}[x]/\mathbb{K}[x]d \cong \mathbb{K}[x]/\mathbb{K}[x]p_1^{m_1} \oplus \mathbb{K}[x]/\mathbb{K}[x]p_2^{m_2} \oplus \cdots \oplus \mathbb{K}[x]/\mathbb{K}[x]p_r^{m_r}
$$

such that the p_i are irreducible polynomials in $\mathbb{K}[x]$ which are pairwise non-associates. (This is an application of the Chinese Remainder Theorem.)

By the cyclic decomposition theorem, *any* finitely generated $\mathbb{K}[x]$-module V_α similarly has a decomposition

$$
V_\alpha \cong \mathbb{K}[x]/\mathbb{K}[x]p_1^{m_1} \oplus \mathbb{K}[x]/\mathbb{K}[x]p_2^{m_2} \oplus \cdots \oplus \mathbb{K}[x]/\mathbb{K}[x]p_s^{m_s}
$$

with the p_i irreducible, $m_i \geq 1$ (but potentially with repeated terms).

Then if $V_\alpha \cong \mathbb{K}[x]/\mathbb{K}[x](x - \lambda)^m$, for $d = (x - \lambda)^m$ we can choose the basis

$$\mathcal{B} = \{(x - \lambda)^{m-i} + \mathbb{K}[x]d \mid 0 \leq i \leq m\}$$

for V_α and see that with respect to this basis we have

$$J_m(\lambda) = \begin{pmatrix} \lambda & 1 & 0 & \cdots & 0 & 0 \\ 0 & \lambda & 1 & \cdots & 0 & 0 \\ 0 & 0 & \lambda & \cdots & 0 & 0 \\ \vdots & \vdots & \vdots & & \vdots & \vdots \\ 0 & 0 & 0 & \cdots & \lambda & 1 \\ 0 & 0 & 0 & \cdots & 0 & \lambda \end{pmatrix}$$

Such a matrix is called a *Jordan block*.

By repeating this process for the decomposition of V_α above, we have that in the general case, there exists a basis for V_α such that the matrix of α is

$$J(\alpha) \stackrel{\text{def}}{=} \begin{pmatrix} J_{m_1}(\lambda_1) & 0 & \cdots & 0 \\ 0 & J_{m_2}(\lambda_2) & \cdots & 0 \\ \vdots & \vdots & & \vdots \\ 0 & 0 & \cdots & J_{m_s}(\lambda_s) \end{pmatrix}$$

We say that $J(\alpha)$ is the *Jordan normal form* of α (sometimes also called the Jordan canonical form).

Similarly, SageMath can find the Jordan normal form provided that all the eigenvalues belong to the base field. Have a play with

```
1  M=random_matrix(QQ,3,3);
2  M.jordan_form()
```

and see how often you obtain sensible output. The first command generates a random 3×3 matrix with entries in $\mathbb{Q}$ whose numerators and denominators are bounded above by 2; this cuts down the complexity somewhat. If you want to manually set the bounds to be something else you can:

```
1  M=random_matrix(QQ,3,3,num_bound=3,den_bound=5);
2  M.jordan_form()
```

Helpfully, SageMath has a function allowing you to specify the input to construct $J_m(\lambda)$ and to take the direct sum of these and so an arbitrary matrix in Jordan normal form:

```
1  M=jordan_block(7,3);
```

```
2  N=jordan_block(-2,4);
3  P=block_diagonal_matrix(M,N);  P
```

If you prefer, you can disable the visual subdivision into blocks:

```
3  P=block_diagonal_matrix(M,N,subdivide=False);  P
```

You can then ask for e.g. the minimal polynomial, or indeed whatever else you wish:

```
4  P.minimal_polynomial().factor()
```

which is $(x - 7)^3(x + 2)^4$, of course.

Problem 28. Create some examples of matrices with different rational canonical and Jordan normal forms.

5.2   Modules for groups

First, we will show how we can translate between groups and algebras in a way that enables us to use all of the above theory for groups.

5.2.1   The group algebra and group of units functors

We are going to describe two recipes, one for turning a group into an algebra and the other an algebra into a group. (Strictly speaking, this isn't correct but the formal statements coming soon will make more sense if we have this as the rough idea in our heads.)

In this section, we will simplify slightly and all our algebras will be $\mathbb{K}$-algebras over a field, rather than just a commutative ring.

The easier direction is getting a group from an algebra, as follows.

Definition 5.2.1. Let $A = (A, m, u)$ be an algebra. The *group of units* of A, $A^\times$, is the group with underlying set

$$A^\times = \{a \in A \mid \exists\, b \in A \text{ such that } ab = 1_A = ba\}$$

(i.e. the set of elements of A having a two-sided multiplicative inverse) and group operation $m|_{A^\times \times A^\times}$ (i.e. the algebra multiplication, restricted to elements of $A^\times$).

That this is a group is straightforward: associativity is inherited from m, $1_A = u(1_\mathbb{K})$ is the identity element and we have inverses by construction. In fact, this construction respects homomorphisms:

Proposition 5.2.2. *There is a functor* $-^\times \colon \mathbf{Alg}_\mathbb{K} \to \mathbf{Grp}$ *given on objects by* $-^\times(A) = A^\times$ *and on morphisms by* $-^\times(f) = f|_{A^\times}$ *for* $f \colon A \to B$.

Proof. We first check that $-^\times$ is well-defined: this follows since if $f\colon A \to B$ and $a \in A^\times$, $f(a) \in B^\times$ because f is an algebra map. In more detail, we may apply f to $ab = 1_A = ba$ to obtain

$$f(a)f(b) = f(ab) = f(1_A) = 1_B = f(1_A) = f(ba) = f(b)f(a)$$

as required.

Clearly $-^\times(\mathrm{id}_A) = \mathrm{id}_A|_{A^\times} = \mathrm{id}_{A^\times}$ so it remains to check composition. Let $f\colon A \to B$ and $g\colon B \to C$. Since $(g \circ f)|_{A^\times} = g|_{B^\times} \circ f|_{A^\times}$, $-^\times$ is a functor. $\qquad\square$

To produce an algebra from a group, we first recall how we can canonically produce a vector space from a set. Well, we know from Section 2.4 how to do this: take the $\mathbb{K}$-linear span of the set freely. Then the set we started with is (by construction) a basis for the vector space.

Let $\mathbb{K}$ be a field and S a set. Recall that $\mathbb{K}[S]$ is the free vector space over S.

In the spirit of Section 2.4, we can phrase this in terms of a functor $\mathbb{K}[-]\colon \mathbf{Set} \to \mathbf{BVect}_\mathbb{K}$ from sets to the category of *based* vector spaces whose objects are pairs $(V, \mathcal{B})$ of a vector space V and a basis $\mathcal{B}$ for V. The reason for using based vector spaces is that there is then a natural *forgetful* functor $b\colon \mathbf{BVect}_\mathbb{K} \to \mathbf{Set}$ with $b(V, \mathcal{B}) = \mathcal{B}$, which has a special relationship with $\mathbb{K}[-]$, namely forming an *adjoint pair* (as per 2.6).

To do this carefully we would need to check how functions transform under $\mathbb{K}[-]$, but instead we just note that the fact that linear maps are determined by their values on a basis is precisely what is needed; we need this fact shortly so let us give the construction a name.

Definition 5.2.3. Let $f\colon S \to T$ be a function. We define $f^{\mathrm{lin}}\colon \mathbb{K}[S] \to \mathbb{K}[T]$ to be the unique $\mathbb{K}$-linear extension of f.

The next step is to show that if G is a group and not just a set, then $\mathbb{K}[G]$ can be made into an algebra. We only have one bit of unused information—the group operation, which we know is associative—so to do something natural and general, we will have to use this to define multiplication. (We said $\mathbb{K}[G]$ is a vector space so it has the addition coming from that structure.)

Proposition 5.2.4. *Let $G = (G, \cdot)$ be a group. Then the maps*

$$m_G\colon \mathbb{K}[G] \otimes \mathbb{K}[G] \to \mathbb{K}[G], \ m_G(g \otimes h) = g \cdot h$$

and

$$u_G\colon \mathbb{K} \to \mathbb{K}[G], \ u_G(\lambda) = \lambda e_G$$

give $\mathbb{K}[G]$ the structure of a $\mathbb{K}$-algebra.

Notice that we have been lazy in defining m_G, only giving its values on a basis. As we will want it later, let us write the most general version.

First, recall that elements of $\mathbb{K}[G]$ have the form $\sum_{g \in G} \alpha_g g$ with $\alpha_g \in \mathbb{K}$ (where if G is infinite, we insist that all but finitely many of the α_g are zero). We then have

$$m_G \left(\left(\sum_{g \in G} \alpha_g g \right) \otimes \left(\sum_{h \in G} \beta_h h \right) \right) = \sum_{g,h \in G} \alpha_g \beta_h (g \cdot h) = \sum_{k \in G} \left(\sum_{\substack{g,h \in G \\ g \cdot h = k}} \alpha_g \beta_h \right) k$$

However, working on the basis will suffice for this proof, as long as we are happy to accept that multiplication on the basis can be uniquely extended to linear combinations in such a way that distributivity holds (reader, it can).

Proof. Associativity follows from associativity in G:

$$
\begin{aligned}
(m_G \circ (m_G \otimes \mathrm{id}))(g \otimes h \otimes k) &= m_G((g \cdot h) \otimes k) \\
&= (g \cdot h) \cdot k \\
&= g \cdot (h \cdot k) \\
&= m_G(g \otimes (h \cdot k)) \\
&= (m_G \circ (\mathrm{id} \otimes m_G))(g \otimes h \otimes k).
\end{aligned}
$$

For the unitarity,

$$
\begin{aligned}
(m_G \circ (u_G \otimes \mathrm{id}))(\lambda \otimes g) &= m_G(\lambda e_G \otimes g) \\
&= \lambda m_G(e_G \otimes g) \\
&= \lambda (e_G \cdot g) \\
&= \lambda g
\end{aligned}
$$

and similarly for $m_G \circ (\mathrm{id} \otimes u_G)$. $\square$

Definition 5.2.5. Let G be a group and $\mathbb{K}$ a field. Then $\mathbb{K}[G] = (\mathbb{K}[G], m_G, u_G)$ is called the *group algebra* of G over $\mathbb{K}$. The dimension of $\mathbb{K}[G]$ is $|G|$.

Example 5.2.6. Let us do an example to see how we do concrete calculations in a group algebra. We take our group to be $C_3 = \langle a \mid a^3 \rangle$ and let us take $\mathbb{K} = \mathbb{C}$.

The defining basis $\mathcal{B}$ for $\mathbb{C}[C_3]$ is the set of elements of C_3, so $\mathcal{B} = \{e, a, a^2\}$; as we said, $\dim \mathbb{C}[C_3] = |C_3| = 3$. So a typical element of $\mathbb{C}[C_3]$ looks like $\alpha e + \beta a + \gamma a^2$ with $\alpha, \beta, \gamma \in \mathbb{C}$, e.g. $3e - \frac{1}{2}a + ia^2$. (If we were working over $\mathbb{R}$, the coefficients α, β, γ would need to be real, e.g. $3e - \frac{1}{2}a + \sqrt{2}a^2$, and you can construct similar examples for $\mathbb{Q}$ or other fields.)

To multiply elements of the group algebra, we are supposed to use the group operation. In C_3, that means remembering that $a^3 = e$.

So for example,

$$(e + 2a)(3e + a^2) = 3e^2 + ea^2 + 6ae + 2a^3$$
$$= 3e + a^2 + 6a + 2e$$
$$= 5e + 6a + a^2$$

Proposition 5.2.7. *There is a functor* $\mathbb{K}[-]\colon \mathbf{Grp} \to \mathbf{Alg}_{\mathbb{K}}$ *given on objects by* $\mathbb{K}[-](G) = \mathbb{K}[G]$ *and on morphisms by* $\mathbb{K}[-](f) = f^{\mathrm{lin}}$.

Proof. We need to show that if $f\colon G \to H$ is a group homomorphism, then $f^{\mathrm{lin}}\colon \mathbb{K}[G] \to \mathbb{K}[H]$ is an algebra homomorphism: we have linearity by construction and

$$f^{\mathrm{lin}}(gh) = f(g \cdot_G h) = f(g) \cdot_H f(h) = f^{\mathrm{lin}}(g)f^{\mathrm{lin}}(h).$$

Clearly $\mathrm{id}_G^{\mathrm{lin}} = \mathrm{id}_{\mathbb{K}[G]}$ since

$$\mathrm{id}_G^{\mathrm{lin}}(g) = \mathrm{id}_G(g) = g = \mathrm{id}_{\mathbb{K}[G]}(g)$$

and $G \subseteq \mathbb{K}[G]$ is a basis. We can also check compositionality on the basis. $\square$

Now we have the results we have been aiming for.

Theorem 5.2.8. *The functors* $\mathbb{K}[-]\colon \mathbf{Grp} \rightleftarrows \mathbf{Alg}_{\mathbb{K}}\colon -^{\times}$ *form an adjoint pair.*

Proof. The claim is that for any $G \in \mathbf{Grp}$ and $A \in \mathbf{Alg}_{\mathbb{K}}$, we have a bijection of sets

$$\alpha_{G,A}\colon \mathrm{Hom}_{\mathbf{Alg}_{\mathbb{K}}}(\mathbb{K}[G], A) \to \mathrm{Hom}_{\mathbf{Grp}}(G, A^{\times})$$

that is functorial in G and A, i.e.

(a) for $f\colon G \to H$,

$$
\begin{array}{ccc}
\mathrm{Hom}_{\mathbf{Alg}_{\mathbb{K}}}(\mathbb{K}[G], A) & \xrightarrow{\alpha_{G,A}} & \mathrm{Hom}_{\mathbf{Grp}}(G, A^{\times}) \\
{\scriptstyle -\circ\mathbb{K}[f]}\big\uparrow & & \big\uparrow{\scriptstyle -\circ f} \\
\mathrm{Hom}_{\mathbf{Alg}_{\mathbb{K}}}(\mathbb{K}[H], A) & \xrightarrow[\alpha_{H,A}]{} & \mathrm{Hom}_{\mathbf{Grp}}(H, A^{\times})
\end{array}
$$

(b) for $p\colon A \to B$,

$$
\begin{array}{ccc}
\mathrm{Hom}_{\mathbf{Alg}_{\mathbb{K}}}(\mathbb{K}[G], A) & \xrightarrow{\alpha_{G,A}} & \mathrm{Hom}_{\mathbf{Grp}}(G, A^{\times}) \\
{\scriptstyle p\circ-}\big\downarrow & & \big\downarrow{\scriptstyle p^{\times}\circ-} \\
\mathrm{Hom}_{\mathbf{Alg}_{\mathbb{K}}}(\mathbb{K}[G], B) & \xrightarrow[\alpha_{G,B}]{} & \mathrm{Hom}_{\mathbf{Grp}}(G, B^{\times})
\end{array}
$$

The key idea is to try to define $\alpha_{G,A}(f\colon \mathbb{K}[G] \to A) = f|_G$. To help keep track of things we will write this slightly differently. Recall that if $X \subseteq Y$ we have the injective function $\iota_X \colon X \to Y$, $\iota_X(x) = x$ and then if $f\colon Y \to Z$ is a function, $f|_X = f \circ \iota_X$.

However, $f \circ \iota_G \colon G \to A$ has the wrong codomain—A rather than $A^\times$. Fortunately(!), $G \subseteq \mathbb{K}[G]^\times$ (and in fact, G is a subgroup[1] of $\mathbb{K}[G]^\times$) so $f(G) \subseteq A^\times$, since f is an algebra homomorphism. So we can restrict the codomain of $f \circ \iota_G$ to $A^\times$.

That is, define

$$\alpha_{G,A}(f) = (f \circ \iota_G)|^{A^\times} \colon G \to A^\times.$$

Then ι_G and f being algebra homomorphisms and the definition of m_G imply that $\alpha_{G,A}(f) \in \mathrm{Hom}_{\mathbf{Grp}}(G, A^\times)$.

We claim that $\alpha_{G,A}$ is a bijection. First assume $\alpha_{G,A}(f) = \alpha_{G,A}(f')$ for $f, f' \in \mathrm{Hom}_{\mathbf{Alg}_\mathbb{K}}(\mathbb{K}[G], A)$. Then $(f \circ \iota_G)|^{A^\times} = (f' \circ \iota_G)|^{A^\times}$, i.e. for all $g \in G$, $f(g) = f'(g)$. Since f and f' then coincide on a basis for $\mathbb{K}[G]$, they are equal. So $\alpha_{G,A}$ is injective.

Now, if $h\colon G \to A^\times$, we have $\mathbb{K}[h]\colon \mathbb{K}[G] \to \mathbb{K}[A^\times]$. But $\mathbb{K}[A^\times] \leq A$ since $A^\times \subseteq A$ and A is a vector space, so $\mathbb{K}[A^\times] = \mathrm{span}_\mathbb{K}(A^\times) \leq A$ (as vector spaces). Let ι_A denote the corresponding map $\iota_A \colon \mathbb{K}[A^\times] \to A$. We claim that $\alpha_{G,A}(\iota_A \circ \mathbb{K}[h]) = h$: we have that for all $g \in G$,

$$\alpha_{G,A}(\iota_A \circ \mathbb{K}[h])(g) = \iota_A \circ \mathbb{K}[h]|_G(g) = \iota_A(h(g)) = h(g),$$

as required. So $\alpha_{G,A}$ is surjective and hence bijective.

Equivalently, let

$$\beta_{G,A}\colon \mathrm{Hom}_{\mathbf{Grp}}(G, A^\times) \to \mathrm{Hom}_{\mathbf{Alg}_\mathbb{K}}(\mathbb{K}[G], A), \ \beta_{G,A}(f) = \iota_A \circ \mathbb{K}[f].$$

Then one may check that $\alpha_{G,A}$ and $\beta_{G,A}$ are inverse to each other.

For (a), we compute. for all $y \in G$, $f\colon G \to H$ and $h \in \mathrm{Hom}_{\mathbf{Alg}_\mathbb{K}}(\mathbb{K}[H], A)$,

$$\begin{aligned}
\alpha_{G,A}(- \circ \mathbb{K}[f])(h)(g) &= \alpha_{G,A}(h \circ \mathbb{K}[f])(g) \\
&= (h \circ \mathbb{K}[f])|_G(g) \\
&= (h \circ f)(g)
\end{aligned}$$

and

$$\begin{aligned}
(- \circ f)(\alpha_{H,A}(h))(g) &= (- \circ f)(h|_G(g)) \\
&= (- \circ f)(h(g)) \\
&= (h \circ f)(g).
\end{aligned}$$

---

[1] There is very deep mathematics around here. A long-standing conjecture of Higman from 1940 (popularized by Kaplansky and so known—erroneously—as Kaplansky's unit conjecture) asserted that $\mathbb{K}[G]^\times = \mathbb{K} \times G$ for G torsion-free. However, a counterexample was given by Gardam in 2021.

For (b), for all $g \in G$, $p: A \to B$ and $f: \mathrm{Hom}_{\mathbf{Alg}_{\mathbb{K}}}(\mathbb{K}[G], A)$,

$$\alpha_{G,B}((p \circ -)(f))(g) = \alpha_{G,B}(p \circ f)(g)$$
$$= (p \circ f)|_G(g)$$
$$= (p \circ f)(g)$$

and

$$(p^\times \circ -)(\alpha_{G,A}(f))(g) = (p^\times \circ f|_G)(g)$$
$$= (p \circ f)(g)$$

as required.   □

Corollary 5.2.9. *For any $G \in \mathbf{Grp}$ and $V \in \mathbf{Vect}_{\mathbb{K}}$, there is a natural bijection*

$$\mathrm{Hom}_{\mathbf{Alg}_{\mathbb{K}}}(\mathbb{K}[G], \mathrm{End}_{\mathbb{K}}(V)) \xrightarrow{\cong} \mathrm{Hom}_{\mathbf{Grp}}(G, \mathrm{GL}(V)).$$

Proof. Simply recall that $\mathrm{End}_{\mathbb{K}}(V)^\times = \mathrm{GL}(V)$.   □

The left-hand side here is (by definition) $\mathrm{Rep}_{\mathbb{K}}(\mathbb{K}[G], V)$ and the right-hand side is (again by definition) $\mathrm{Rep}_{\mathbb{K}}(G, V)$. The above bijection between these exists for any V and is natural, i.e. if V and W are two different vector spaces, the corresponding bijections are compatible with homomorphisms between V and W. So these bijections "fit together" to give an isomorphism of categories between $\mathrm{Rep}_{\mathbb{K}}(\mathbb{K}[G])$ and $\mathrm{Rep}_{\mathbb{K}}(G)$.

Since we have also seen in Theorem 4.1.25 that, as categories, $\mathrm{Rep}_{\mathbb{K}}(\mathbb{K}[G]) \cong \mathbb{K}[G]\text{-}\mathbf{Mod}$, we have

> Linear representations of G correspond to $\mathbb{K}[G]$-modules.

Example 5.2.10. Let G be a group acting on a set Ω via $\triangleright$. Then we can *linearize* the action to a linear representation: let $\mathbb{K}[\Omega]$ be the $\mathbb{K}$-vector space with basis Ω and define $\rho: G \to \mathrm{GL}(\mathbb{K}[\Omega])$ by $\rho(g)(x) = g \triangleright x$, extended linearly, i.e. $\rho(g) = (g \triangleright -)^{\mathrm{lin}}$.

This representation corresponds to the $\mathbb{K}[G]$-module $(\mathbb{K}[\Omega], \rho^{\mathrm{lin}})$. Following through the definitions, we see that the module structure map here is really just the linearization of the original action, $\triangleright$.

This module is known as the *permutation module* associated to the action of G on Ω. An important special case is $\Omega = G$ with action being left multiplication, corresponding to the regular representation.

5.2.2 Semisimplicity

Recall that any algebra has a *centre*,

$$Z(A) = \{z \in A \mid za = az \; \forall \, a \in A\},$$

consisting of all the elements z of A that commute with every element of A.

Problem 29. Show that $Z(A)$ is a subalgebra of A.

When G is a finite group, the centre of its group algebra has a particularly nice description, as follows. First, recall that a group G is partitioned into its *conjugacy classes*: these are precisely the orbits for the action of G on itself via $g \triangleright h = ghg^{-1}$, or more explicitly, the conjugacy class containing $h \in G$ is

$$C(h) = \{ghg^{-1} \mid g \in G\}.$$

Proposition 5.2.11. *Let G be a finite group and enumerate the conjugacy classes of G as $\{C_i \mid 1 \le i \le r\}$. Define $\overline{C}_i \overset{\mathrm{def}}{=} \sum_{g \in C_i} g$, the class sum, as an element of $\mathbb{K}[G]$. Then $\{\overline{C}_i \mid 1 \le i \le r\}$ is a basis for $Z(\mathbb{K}[G])$, which in particular has dimension r.*

Proof. We first show that the class sums are central. Fixing $g \in C_i$, we may choose elements $y_j \in G$ ($1 \le j \le m = |C_i|$) so that we may write $C_i = \{y_j^{-1} g y_j \mid 1 \le j \le m\}$, and hence $\overline{C}_i = \sum_{j=1}^{m} y_j^{-1} g y_j$. Then for all $h \in G$,

$$h^{-1} \overline{C}_i h = \sum_{j=1}^{m} h^{-1} y_j^{-1} g y_j h = \sum_{j=1}^{m} (y_j h)^{-1} g(y_j h) = \overline{C}_i$$

since C_i is a conjugacy class, so as j runs from 1 to m, $(y_j h)^{-1} g(y_j h)$ will run through C_i, since $y_j^{-1} g y_j$ does. So $\overline{C}_i h = h \overline{C}_i$ and $\overline{C}_i$ is central in $\mathbb{K}[G]$: by linearity, it suffices for $\overline{C}_i$ to commute with elements of G.

Now $\{\overline{C}_i \mid 1 \le i \le r\}$ is a linearly independent set, by the disjointness of conjugacy classes. This set also spans $Z(\mathbb{K}[G])$: let $z = \sum_{g \in G} \lambda_g g \in \mathbb{K}[G]$. Then for any $h \in G$, $h^{-1} z h = z$ so $\sum_{g \in G} \lambda_g h^{-1} g h = \sum_{g \in G} \lambda_g g$. Hence for every $h \in G$ the coefficient λ_g of g in z is the same as $\lambda_{h^{-1} gh}$, i.e. that of $h^{-1} gh$. So the function $g \to \lambda_g$ is constant on conjugacy classes and so $z = \sum_{i=1}^{r} \lambda_i \overline{C}_i$, with λ_i the coefficient of any $g \in C_i$. $\qquad \square$

Example 5.2.12. A basis for $Z(\mathbb{K}[S_3])$ is

$$\{\iota, (1\,2) + (1\,3) + (2\,3), (1\,2\,3) + (1\,3\,2)\}$$

and $\dim Z(\mathbb{K}[S_3]) = 3$ (compared to $\dim \mathbb{K}[S_3] = |S_3| = 6$).

SageMath can help here:

```
1   G = SymmetricGroup(3)
2   G.conjugacy_classes()
```

We need to interpret the output appropriately, but we can immediately see that SageMath has told us that there are three conjugacy classes.

We now return to the special case of $\mathbb{K}[G]$, the group algebra of a finite group. By our previous remarks about semisimple algebras, the following theorem tells us (essentially) everything we need to know.

Recall that for any field $\mathbb{K}$, there is a (non-zero) ring homomorphism $u\colon \mathbb{Z} \to \mathbb{K}$, $u(1_\mathbb{Z}) = 1_\mathbb{K}$ (extended linearly: $u(n1_\mathbb{Z}) = n1_\mathbb{K}$). The kernel of u is a proper ideal of $\mathbb{Z}$, which is a principal ideal domain, and therefore either $\operatorname{Ker} u = 0$ or $\operatorname{Ker} u = n\mathbb{Z}$ for some $n \in \mathbb{Z}$.

Now since $\mathbb{K}$ is a field, $\operatorname{Im} u$ is an integral domain and $\operatorname{Im} u \cong \mathbb{Z}/n\mathbb{Z}$ by the First Isomorphism Theorem for Rings. But $\mathbb{Z}/n\mathbb{Z}$ is an integral domain if and only if n is prime. So $\operatorname{Ker} u = 0$ or $\operatorname{Ker} u = p\mathbb{Z}$ for some prime p.

If $\operatorname{Ker} u = 0$, we say $\mathbb{K}$ has *characteristic zero*; if $\operatorname{Ker} u = p\mathbb{Z}$ with p prime, we say $\mathbb{K}$ has *characteristic p*. We denote the characteristic of $\mathbb{K}$ by char $\mathbb{K}$.

Theorem 5.2.13 (Maschke). *Let G be a finite group. Then the group algebra $\mathbb{K}[G]$ is semisimple if and only if* char $\mathbb{K}$ *does not divide* $|G|$.

Note that 0 never divides $|G|$, i.e. over fields of characteristic zero—such as $\mathbb{R}$ and $\mathbb{C}$—the group algebra of a finite group is always semisimple.

Before we prove Maschke's theorem, we need a lemma.

Lemma 5.2.14. *Let G be a finite group and $\mathbb{K}$ a field. Let M and N be $\mathbb{K}[G]$-modules and let $f\colon M \to N$ be a $\mathbb{K}$-linear map. Then*

$$T(f)\colon M \to N, \quad T(f)(m) = \sum_{g \in G} g \triangleright (f(g^{-1} \triangleright m))$$

is a $\mathbb{K}[G]$-module homomorphism.

Proof. Exercise. $\qquad\qquad\qquad\qquad\qquad\qquad\qquad\qquad\qquad\qquad\qquad\qquad\square$

Proof of Maschke's theorem. Assume char $\mathbb{K} \nmid |G|$. Let M be a submodule of $_{\mathbb{K}[G]}\mathbb{K}[G]$ and let N be a *vector space* complement to M, i.e. we have $_{\mathbb{K}[G]}\mathbb{K}[G] \cong_{\text{v.s.}} M \oplus N$. Let $f\colon {}_{\mathbb{K}[G]}\mathbb{K}[G] \to M$ be the *linear* map with kernel N. Then char $\mathbb{K} \nmid |G|$ implies $|G| \equiv |G|1_\mathbb{K}$ is invertible in $\mathbb{K}$ and so

$$\gamma\colon {}_{\mathbb{K}[G]}\mathbb{K}[G] \twoheadrightarrow M, \quad \gamma \overset{\text{def}}{=} \frac{1}{|G|} T(f)$$

is well-defined. By the lemma, γ is a $\mathbb{K}[G]$-module homomorphism. Let $K = \operatorname{Ker} \gamma$.

We claim that $i\colon M \hookrightarrow {}_{\mathbb{K}[G]}\mathbb{K}[G]$ splits γ (in the sense of 4.2.8). Since $f(m) = m$ for all $m \in M$, $f(g^{-1} \triangleright m) = g^{-1} \triangleright m$ and hence $g \triangleright (f(g^{-1} \triangleright m)) = g \triangleright g^{-1} \triangleright m = m$ for all $m \in M$. Hence

$$(\gamma \circ i)(m) = \frac{1}{|G|} \sum_{g \in G} m = m$$

as required. Then $_{\mathbb{K}[G]}\mathbb{K}[G] \cong M \oplus K$ as $\mathbb{K}[G]$-modules, so by Lemma 4.3.5 $_{\mathbb{K}[G]}\mathbb{K}[G]$ is semisimple, hence $\mathbb{K}[G]$ is semisimple.

Now assume $\mathbb{K}[G]$ is semisimple. Consider $w \stackrel{\text{def}}{=} \sum_{g \in G} g \in \mathbb{K}[G]$. Then $hw = w$ for all $h \in G$ so $\mathbb{K}w \leq {}_{\mathbb{K}[G]}\mathbb{K}[G]$ is a 1-dimensional submodule of ${}_{\mathbb{K}[G]}\mathbb{K}[G]$. Since $\mathbb{K}[G]$ is semisimple, there exists C such that as modules we have ${}_{\mathbb{K}[G]}\mathbb{K}[G] \cong \mathbb{K}w \oplus C$.

Then there exists $c \in C$ such that $1_{\mathbb{K}[G]} = \lambda w + c$ for $\lambda \in \mathbb{K}$, $c \in C$. We have $\lambda \neq 0$, else ${}_{\mathbb{K}[G]}\mathbb{K}[G] = \mathbb{K}[G]C \subseteq C \neq {}_{\mathbb{K}[G]}\mathbb{K}[G]$, a contradiction.

We see that $w^2 = |G|w$ and

$$w = w 1_{\mathbb{K}[G]} = w(\lambda w + c) = \lambda |G| w + wc.$$

Now $wc \in C$ so $w - \lambda |G| w \in C$ and $w - \lambda |G| w = (1 - \lambda |G|) w \in \mathbb{K}w$. But $\mathbb{K}w \cap C = \{0\}$ and $w \neq 0$, $\lambda \neq 0$, so we conclude that $1 - \lambda |G| = 0$ in $\mathbb{K}$, i.e. $|G| = \lambda^{-1} \neq 0$ in $\mathbb{K}$. That is, char $\mathbb{K} \nmid |G|$. $\qquad\square$

Remark 5.2.15. With a little more technology, one can prove the stronger form of Maschke's theorem: $\mathbb{K}[G]$ is semisimple if and only if G is finite and char $\mathbb{K}$ does not divide $|G|$. See, for example, [Lor, Section 3.4].

Note that the finiteness of G is crucial: the conclusion is not necessarily true if G is not finite. See Exercise 5.19 for an example of what can go wrong.

Consequently,

Theorem 5.2.16. *Let G be a finite group and $\mathbb{K}$ an algebraically closed field of characteristic zero. Then*

(i) $\mathbb{K}[G] \cong \prod_{i=1}^{r} M_{n_i}(\mathbb{K})$ *as algebras for some $n_i \in \mathbb{N}$*

(ii) *$\mathbb{K}[G]$ has r pairwise non-isomorphic simple modules $\{S_i \mid 1 \leq i \leq r\}$ with dimension $\dim S_i = n_i$, and*

$$_{\mathbb{K}[G]}\mathbb{K}[G] \cong \bigoplus_{i=1}^{r} S_i^{\oplus n_i}$$

(iii) *$|G| = \sum_{i=1}^{r} n_i^2$*

(iv) *there exists a complete set of central orthogonal idempotents $\{e_i \mid 1 \leq i \leq r\}$ in $\mathbb{K}[G]$*

　　(That is, there are elements e_i satisfying: $e_i^2 = e_i$ (idempotent), $e_i e_j = 0$ for $i \neq j$ (orthogonal), $e_i \in Z(\mathbb{K}[G])$ (central) and $\sum_i e_i = 1$ (complete set).)

(v) *$r = \dim Z(\mathbb{K}[G])$ is equal to the number of conjugacy classes of G*

(vi) *G is Abelian if and only if $\dim S_i = 1$ for all i if and only if $|G| = r$.*

Proof.

(i) The assumptions of the theorem allow us to apply Maschke's theorem, so that $\mathbb{K}[G]$ is semisimple. Then the claim follows from the Artin–Wedderburn theorem (specifically, from Corollary 4.5.10).

(ii) The matrix algebra $M_{n_i}(\mathbb{K})$ has exactly one simple module, the natural module, isomorphic to $\mathbb{K}^{\oplus n_i}$, of dimension n_i. Set $S_i = \mathbb{K}^{\oplus n_i}$. It is straightforward to check that all simples for $\prod_{i=1}^{r} M_{n_i}(\mathbb{K})$ are of the form

$$0 \times 0 \times \cdots \times 0 \times S_i \times 0 \times \cdots \times 0.$$

Furthermore, $M_{n_i}(\mathbb{K})$ is a simple algebra of dimension n_i^2 so

$$_{M_{n_i}(\mathbb{K})} M_{n_i}(\mathbb{K}) \cong S_i^{\oplus n_i},$$

from which the last claim follows.

(iii) This follows by comparing dimensions in the isomorphism in part (ii).

(iv) Define $B_i = S_i^{\oplus n_i}$, so $_{\mathbb{K}[G]}\mathbb{K}[G] \cong \bigoplus_{i=1}^{r} B_i$. Then there exist elements $\{e_i \in B_i \mid 1 \le i \le r\}$ such that $1_{\mathbb{K}[G]} = \sum_{i=1}^{r} e_i$.

Consider e_i and e_j. Since B_j is a submodule, $e_i e_j \in B_j$. On the other hand,

$$e_i = e_i 1_{\mathbb{K}[G]} = e_i(e_1 + \cdots + e_r) = e_i e_1 + \cdots + e_i e_r.$$

Then since $e_i \in B_i$, $e_i e_j \in B_j$ and the sum $_{\mathbb{K}[G]}\mathbb{K}[G] \cong \bigoplus_{i=1}^{r} B_i$ is direct, we must have $e_i e_j = 0$ for $i \ne j$ and hence $e_i = e_i e_i$. That is, e_i is an idempotent and the set $\{e_i\}$ is a complete set of orthogonal idempotents.

For $b \in B_i$, $b 1_{\mathbb{K}[G]} = b = 1_{\mathbb{K}[G]} b$, $1_{\mathbb{K}[G]} = \sum_{i=1}^{r} e_i$ and the orthogonality of the idempotents together imply that $b e_i = b = e_i b$ so that e_i is the multiplicative identity of B_i, hence identified with the identity matrix in $M_{n_i}(\mathbb{K})$ in the algebra decomposition $\mathbb{K}[G] \cong \prod M_{n_i}(\mathbb{K})$. The product structure implies that each e_i commutes with the other factors $M_{n_j}(\mathbb{K})$ for $j \ne i$, since $e_i b = 0 = b e_i$ for $b \in B_j$, $j \ne i$, and hence the e_i are central.

(v) Now the e_i span a subspace of $Z(\mathbb{K}[G])$ of dimension r. By Schur's lemma, any $z \in Z(\mathbb{K}[G])$ acts on S_i by a scalar λ_i and $z e_i = \lambda_i e_i$. Then

$$z = z 1_{\mathbb{K}[G]} = z \left(\sum_{i=1}^{r} e_i \right) = \sum_{i=1}^{r} \lambda_i e_i \in \mathrm{span}_{\mathbb{K}}\{e_i \mid 1 \le i \le r\}.$$

So $\dim Z(\mathbb{K}[G]) \le r$ and hence $\dim Z(\mathbb{K}[G]) = r$ (and each e_i spans $Z(B_i)$). By Proposition 5.2.11, this is the number of conjugacy classes.

(vi) Note first that $M_{n_i}(\mathbb{K})$ is commutative if and only if $n_i = 1$. So if G is Abelian, and hence $\mathbb{K}[G]$ is commutative, we must have $n_i = 1$ for all i and by (ii), all the simples are 1-dimensional.

If all simples are 1-dimensional, then by (iii) $|G| = r$. Then every conjugacy class of G must be a singleton, which implies G is Abelian.

$\square$

The take-away from this theorem is that for finite groups and over algebraically closed fields of characteristic zero, we have very detailed information on their representation theory.

Let us specialize one step further and look at what this means for the simplest groups, namely finite Abelian groups. To avoid field-theoretic complications, let us take $\mathbb{K} = \mathbb{C}$.

If G is a finite Abelian group, then as shown in Section 5.1, noting that $C_{d_i} \cong \mathbb{Z}/\mathbb{Z}d_i$ we have

$$G \cong C_{d_1} \times C_{d_2} \times \cdots \times C_{d_r}$$

where $C_{d_i} = \left\langle g_i \mid g_i^{d_i} = e \right\rangle$. In particular, if $g \in G$ we can write (by a mild abuse of notation) $g = g_1^{a_1} \cdots g_r^{a_r}$ with $a_i \in \mathbb{Z}$.

Then letting $\zeta_i \in \mathbb{C}$ be a d_ith root of unity and $\underline{\zeta} = (\zeta_1, \ldots, \zeta_r)$, define $V_{\underline{\zeta}}$ to be the 1-dimensional module $V_{\underline{\zeta}} = \mathbb{C}v$ with

$$g_1^{a_1} \cdots g_r^{a_r} \triangleright v = \zeta_1^{a_1} \cdots \zeta_r^{a_r} v.$$

These are 1-dimensional and hence necessarily simple.

There are d_i d_ith roots of unity, so $\prod_{i=1}^{r} d_i$ tuples $\underline{\zeta} = (\zeta_1, \ldots, \zeta_r)$. But $\prod_{i=1}^{r} d_i = |G|$ so we have the right number of (1-dimensional) simple modules. With a little more work, one can check that these are pairwise non-isomorphic and hence are exactly the simple modules.

Example 5.2.17. As an explicit example, consider $G = C_4$. Then our theory tells us that

$$\mathbb{C}[C_4] \cong M_1(\mathbb{C}) \times M_1(\mathbb{C}) \times M_1(\mathbb{C}) \times M_1(\mathbb{C}).$$

Each $M_1(\mathbb{C}) \equiv \mathbb{C}$ has one 1-dimensional simple module, so $\mathbb{C}[C_4]$ has four 1-dimensional simples. These exactly correspond to the eigenspaces associated with the four 4th roots of unity, $1, -1, i$ and $-i$.

The first two of the following exercises are closely related to this.

Problem 30. Let $G = \left\langle a \mid a^4 = e \right\rangle = C_4$ be a cyclic group of order 4 and let M be the $\mathbb{C}[G]$-module of dimension 4 with basis $\mathcal{B} = \{m_1, m_2, m_3, m_4\}$ with

$$a \triangleright m_1 = m_2, \ a \triangleright m_2 = m_3, \ a \triangleright m_3 = m_4, \ a \triangleright m_4 = m_1.$$

(a) Show that $M_1 = \operatorname{span}_{\mathbb{C}}\{m_1 + m_2 + m_3 + m_4\}$ is a 1-dimensional $\mathbb{C}[G]$-submodule of M.

(b) Find three other 1-dimensional $\mathbb{C}[G]$-submodules M_2, M_3 and M_4 of M such that $M = M_1 \oplus M_2 \oplus M_3 \oplus M_4$.

[*Hint: consider vectors of the form* $m_1 + \lambda m_2 + \lambda^2 m_3 + \lambda^3 m_4$ *for suitable values of* λ.]

Problem 31. Let n be an integer $n \geq 2$, let $G = \langle a \mid a^n = e \rangle$ be a cyclic group of order n and let M be the permutation $\mathbb{C}[G]$-module of dimension n with natural basis $\mathcal{B} = \{m_1, \ldots, m_n\}$ defined by

$$a \triangleright m_i = m_{i+1}$$

where we interpret the indices modulo n. Find all the 1-dimensional $\mathbb{C}[G]$-submodules of M.

Problem 32. Let G be a group of order 12 and $\{S_1, \ldots, S_r\}$ be a complete set of non-isomorphic simple $\mathbb{C}[G]$-modules. List all the possible integers r and hence all the sequences of dimensions

$$(\dim S_1, \dim S_2, \ldots, \dim S_r).$$

5.3 Modules for quivers

As for groups, we need to find a way to invoke the theory of modules over algebras to study quiver representations. That is, we need to find an algebra whose module category is the same as the category of quiver representations. We can do this by constructing the *path algebra*. After we see that this does indeed encode the right representation theory, we will work through the fundamental questions: what are the simple, projective and injective modules? When is the category semisimple? By contrast with the group case, the answer to the former question is easier but the answer to the latter is "rarely", so the representation theory we will encounter in this section has some different features.

5.3.1 The path algebra

Let $\mathcal{Q}$ be a quiver with vertex set $\text{Vert}(\mathcal{Q})$, arrows $\mathcal{Q}(v, w)$ (for all pairs $v, w \in \text{Vert}(\mathcal{Q})$). Recall that we have the associated path category (Definition 2.1.3), with objects $\text{Vert}(\mathcal{Q})$ and morphisms $\mathcal{P}(\mathcal{Q})(v, w) = \mathcal{P}(v, w)$ for $v, w \in \text{Vert}(\mathcal{Q})$, where $\mathcal{P}(v, w)$ denotes the collection of finite paths from v to w. To construct the path *algebra*, we "linearize" the path category.

Let $\mathbb{K}$ be a field and $\mathcal{Q}$ a quiver. Recall that we use the notation $\mathbb{K}[S]$ to denote the free $\mathbb{K}$-vector space over S, i.e. the vector space spanned by the elements of S.

Let $\mathbb{K}[\mathcal{Q}]$ be the $\mathbb{K}$-vector space

$$\mathbb{K}[\mathcal{Q}] \stackrel{\text{def}}{=} \bigoplus_{v, w \in \text{Vert}(\mathcal{Q})} \mathbb{K}[\mathcal{P}(v, w)]$$

so that $\mathbb{K}[\mathcal{Q}]$ is the $\mathbb{K}$-vector space spanned by all (finite) paths in $\mathcal{Q}$.

We can make $\mathbb{K}[\mathcal{Q}]$ into an algebra, by equipping it with a multiplication and a unit. To define multiplication, we take a pair of basis elements, i.e. paths $p \in \mathcal{P}(v, w)$ and $q \in \mathcal{P}(x, y)$. Then define

$$m(q, p) = \begin{cases} q \circ p & \text{if defined, i.e. } w = x \\ 0 & \text{otherwise} \end{cases}$$

That is, the multiplication in the path algebra matches the composition of paths in the path category. Then m is defined as a linear map $m \colon \mathbb{K}[\mathcal{Q}] \otimes \mathbb{K}[\mathcal{Q}] \to \mathbb{K}[\mathcal{Q}]$ by extending the above linearly. Let us now just write qp for $m(q, p)$.

Recall that we denote by e_v the trivial path of length 0 from a vertex v to itself.

Lemma 5.3.1. *Let $\mathbb{K}$ be a field and let $\mathcal{Q}$ be a quiver.*

(i) *The trivial path e_v is an idempotent in $\mathbb{K}[\mathcal{Q}]$, i.e. $e_v e_v = e_v$.*

(ii) *The idempotents e_v are orthogonal: if $v \neq w$ then $e_v e_w = 0$.*

(iii) *For any path $p \in \mathcal{P}(v, w)$ we have $e_w p = p = p e_v$.*

(iv) *Let $\mathcal{Q}$ be a finite quiver and set*

$$1_{\mathbb{K}[\mathcal{Q}]} \stackrel{\text{def}}{=} \sum_{v \in \mathrm{Vert}(\mathcal{Q})} e_v$$

For any path $p \in \mathcal{P}(v, w)$ we have $1_{\mathbb{K}[\mathcal{Q}]} p = p = p 1_{\mathbb{K}[\mathcal{Q}]}$.

Proof. Exercise.   □

Remark 5.3.2. Since $1_{\mathbb{K}[\mathcal{Q}]}$ is not a well-defined element of $\mathbb{K}[\mathcal{Q}]$ unless $\mathcal{Q}$ has finitely many vertices and since we want our algebras to be unital, from this point on, *all our quivers will be finite* (usually without further comment). One can handle the infinite case either by being more relaxed about unitarity (the system of orthogonal idempotents $\{e_v\}$ exists in general and can be used in place of a unit for many things) or by working in a slightly larger algebra than $\mathbb{K}[\mathcal{Q}]$ as we have defined it (a *completion*, to be precise, to let us work with formal infinite sums).

Remark 5.3.3. For use later, let us discuss some subspaces of $\mathbb{K}[\mathcal{Q}]$.

Firstly, consider $\mathbb{K}[\mathcal{Q}]e_v$. We claim that this subspace has as basis all paths in $\mathcal{Q}$ starting at v. For if p is a path in $\mathcal{Q}$, $p e_v = 0$ if $t(p) \neq v$ and $p e_v = p$ if $t(p) = v$. So $\mathbb{K}[\mathcal{Q}]e_v$ is spanned by $\{p \mid t(p) = v\}$ and these are linearly independent as this is a subset of the basis of $\mathbb{K}[\mathcal{Q}]$.

Similarly, $e_w \mathbb{K}[\mathcal{Q}]$ has as basis all paths ending at w.

Finally, we see that $e_w \mathbb{K}[\mathcal{Q}]e_v = \mathbb{K}[\mathcal{P}(v, w)]$ has as basis all paths from v to w.

Note too that $\mathbb{K}[\mathcal{Q}]e_v$ is naturally a submodule of the (left) regular module $_{\mathbb{K}[\mathcal{Q}]}\mathbb{K}[\mathcal{Q}]$, since for any path q, $q \triangleright (pe_v) = (qp) \triangleright e_v = (qp)e_v \in \mathbb{K}[\mathcal{Q}]e_v$. Conversely, $e_w\mathbb{K}[\mathcal{Q}]$ is naturally a right module; indeed, it is a submodule of the right regular module $\mathbb{K}[\mathcal{Q}]_{\mathbb{K}[\mathcal{Q}]}$. It is mildly irritating that this is not a left module but that's [life/mathematics]. Indeed, $e_w\mathbb{K}[\mathcal{Q}]e_v$ is not a left or right module.

Now, defining $u \colon \mathbb{K} \to \mathbb{K}[\mathcal{Q}]$, $u(\lambda) = \lambda 1_{\mathbb{K}[\mathcal{Q}]}$, we may see that $(\mathbb{K}[\mathcal{Q}], m, u)$ is a $\mathbb{K}$-algebra. Note that associativity is immediate from the definition of composition of paths in the path category and the previous lemma covers unitarity.

Definition 5.3.4. Let $\mathbb{K}$ be a field and $\mathcal{Q}$ a quiver. The algebra $(\mathbb{K}[\mathcal{Q}], m, u)$ is called the *path algebra* of $\mathcal{Q}$.

Remark 5.3.5. We use the notation $\mathbb{K}[\mathcal{Q}]$ for the path algebra in a very deliberate echo of the use of both $\mathbb{K}[S]$ for free vector spaces and $\mathbb{K}[G]$ for the group algebra. However, note that many authors write $\mathbb{K}\mathcal{Q}$ for the path algebra; to be fair, some also write $\mathbb{K}G$ for the group algebra.

Indeed, the path algebra can be thought of as a free construction, like the group algebra is, and "almost" gives rise to an adjoint pair of functors. Unfortunately there are some technical difficulties, coming from the fact that (unlike the group of units functor) the reverse construction, known as the *Gabriel quiver* of an algebra, is not functorial. There are also some issues with exactly which quivers and exactly which algebras should be allowed. For the interested reader, this story is explained in detail in [IM].

SageMath can construct path algebras by taking the semigroup algebra over the path semigroup, as follows, which we illustrate using the example from Example 1.6.3.

```
1  Q = DiGraph({1:{4:'e14'},2:{4:'e24'},3:{4:'e34'}})
2  PQ = Q.path_semigroup()
3  A = PQ.algebra(QQ)
```

Note that by definition the path algebra is finite-dimensional if and only if the quiver $\mathcal{Q}$ has finitely many paths in it. As soon as $\mathcal{Q}$ contains an oriented cycle c, the path algebra is infinite-dimensional; an infinite-dimensional subspace is spanned by $\{c^n \mid n \in \mathbb{N}\}$. On the other hand, a finite acyclic quiver has a finite-dimensional path algebra. Working out the dimension of the path algebra when this is finite is not as easy as finding the dimension of the group algebra, in general.

That said, computers can sometimes be helpful! Continuing the above,

4   `A.dimension()`

returns 7, since Q has four vertices and three arrows but no paths of higher length.

Since $\mathbb{K}[\mathcal{Q}]$ is a $\mathbb{K}$-algebra, it has a category of modules, $\mathbb{K}[\mathcal{Q}]$-Mod. We now show that this is equivalent to the category $\mathrm{Rep}(\mathcal{Q})$ of linear representations of $\mathcal{Q}$ (Definition 3.3.5).

Proposition 5.3.6. *Let $\mathbb{K}$ be a field and $\mathcal{Q}$ a quiver. There is a functor $\mathcal{M}\colon \mathrm{Rep}(\mathcal{Q}) \to \mathbb{K}[\mathcal{Q}]$-Mod defined on objects $F\colon \mathcal{P}(\mathcal{Q}) \to \mathbb{K}$-Mod by*

$$\mathcal{M}(F) = (\mathcal{M}F \stackrel{\mathrm{def}}{=} \bigoplus_{v \in \mathrm{Vert}(\mathcal{Q})} Fv, \triangleright),$$

with

$$\triangleright\colon \mathbb{K}[\mathcal{Q}] \otimes \mathcal{M}F \to \mathcal{M}F, \ p \triangleright x = \begin{cases} (Fp)(x) & \text{if } x \in Fv, p \in P(v,w) \\ 0 & \text{otherwise} \end{cases}$$

and on morphisms $\alpha\colon F \Rightarrow G$ by

$$\mathcal{M}(\alpha)\colon \mathcal{M}F \to \mathcal{M}G, \ \mathcal{M}(\alpha) = \bigoplus_{v \in \mathrm{Vert}(\mathcal{Q})} \alpha_v$$

Proof. To see that $\mathcal{M}$ is well-defined, we check that $\triangleright$ as stated does give a $\mathbb{K}[\mathcal{Q}]$-module structure, i.e. $q \triangleright (p \triangleright x) = qp \triangleright x$. Consider $p \in P(v,w)$, $q \in P(w,z)$ and $x \in \mathcal{M}F$. If $x \notin Fv$, $p \triangleright x = 0$ and there is nothing to check. If $x \in Fv$,

$$q \triangleright (p \triangleright x) = q \triangleright (Fp)(x) = (Fq)((Fp)(x)) = F(q \circ p)(x) = qp \triangleright x.$$

We also verify that $\mathcal{M}(\alpha)$ is a module homomorphism: for $p \in P(v,w)$ and $x \in Fv$),

$$\mathcal{M}(\alpha)(p \triangleright x) = \left(\bigoplus_v \alpha_v\right)(Fp)(x)$$

$$= (\alpha_w \circ Fp)(x)$$

$$= (Gp \circ \alpha_v)(x)$$

$$= p \triangleright \left(\bigoplus_v \alpha_v\right)(x)$$

$$= p \triangleright \mathcal{M}(\alpha)(x).$$

If $\alpha\colon F \Rightarrow G$ and $\beta\colon G \Rightarrow H$ with components α_v and β_v respectively then functoriality of $\mathcal{M}$ follows from the identity

$$\left(\bigoplus_v \beta_v\right) \circ \left(\bigoplus_v \alpha_v\right) = \bigoplus_v \beta_v \circ \alpha_v$$

and the straightforward check that $\mathcal{M}$ respects identity maps. $\square$

Given $M \in \mathbb{K}[\mathcal{Q}]$-Mod, for $v \in \mathrm{Vert}(\mathcal{Q})$ we define

$$e_v M = \{e_v \triangleright m \mid m \in M\}$$

where e_v is the idempotent in $\mathbb{K}[\mathcal{Q}]$ defined in Lemma 5.3.1. Note that $e_v M$ is a subspace of M but not a submodule. However, as vector spaces, $M = \bigoplus_{v \in \mathrm{Vert}(\mathcal{Q})} e_v M$; this follows from orthogonality of the e_v.

Now if $p \in \mathcal{P}(v, w)$ and $m \in M$, we have that

$$p \triangleright (e_v \triangleright m) = (pe_v) \triangleright m = p \triangleright m = (e_w p) \triangleright m = e_w \triangleright (p \triangleright m) \in e_w M.$$

This allows us to make the definitions in the following.

Proposition 5.3.7. *Let $\mathbb{K}$ be a field and $\mathcal{Q}$ a quiver. There is a functor $\mathcal{R}\colon \mathbb{K}[\mathcal{Q}]$-Mod $\to \mathrm{Rep}(\mathcal{Q})$ defined on objects $M = (M, \triangleright)$ by*

$\mathcal{R}(M)\colon \mathcal{P}(\mathcal{Q}) \to \mathbb{K}$-Mod,

$\quad \mathcal{R}(M)(v) = e_v M$ *for $v \in \mathrm{Vert}(\mathcal{Q})$,*

$\quad \mathcal{R}(M)(p)\colon e_v M \to e_w M$, $\mathcal{R}(M)(p)(e_v \triangleright m) = p \triangleright m$ *for $p \in \mathcal{P}(v, w)$*

and on morphisms $f\colon M \to N$ by

$$\mathcal{R}(M)(f)\colon \mathcal{R}(M) \Rightarrow \mathcal{R}(N), \mathcal{R}(M)(f)_v\colon e_v M \to e_v N, \mathcal{R}(M)(f)_v = f|_{e_v M}$$

Proof. We need to verify that $\mathcal{R}(M)$ is a functor on the path category. For this, we require that for paths $p \in \mathcal{P}(v, w)$ and $q \in \mathcal{P}(w, z)$ we have $\mathcal{R}(M)(q) \circ \mathcal{R}(M)(p) = \mathcal{R}(M)(q \circ p)$. Now for $e_v \triangleright m \in e_v M$,

$$
\begin{aligned}
(\mathcal{R}(M)(q) \circ \mathcal{R}(M)(p))(e_v \triangleright m) &= \mathcal{R}(M)(q)(p \triangleright m) \\
&= q \triangleright (p \triangleright m) \\
&= qp \triangleright m \\
&= \mathcal{R}(M)(q \circ p)(e_v \triangleright m)
\end{aligned}
$$

and $\mathcal{R}(M)(e_v)(e_v \triangleright m) = e_v^2 \triangleright m = e_v \triangleright m$ so $\mathcal{R}(M)(e_v) = \mathrm{id}_{e_v M}$ as required.

Furthermore, f being a module homomorphism implies that $\mathcal{R}(M)(f)$ is a natural transformation. Let $p \in \mathcal{P}(v, w)$. Then

$$
\begin{aligned}
(\mathcal{R}(M)(f)_w \circ \mathcal{R}(M)(p))(e_v \triangleright m) &= \mathcal{R}(M)(f)_w(p \triangleright (e_v \triangleright m)) \\
&= f(p \triangleright (e_v \triangleright m)) \\
&= p \triangleright f(e_v \triangleright m) \\
&= (\mathcal{R}(M)(p) \circ \mathcal{R}(M)(f)_v)(e_v \triangleright m)
\end{aligned}
$$

as required.

For functoriality of $\mathcal{R}$, we require that if $f\colon M \to N$ and $g\colon N \to P$, $\mathcal{R}(M)(g \circ f) = \mathcal{R}(M)(g) \circ \mathcal{R}(M)(f)$. Let $v \in \mathrm{Vert}(\mathcal{Q})$. Then

$$
\begin{aligned}
\mathcal{R}(M)(g \circ f)_v(e_v \rhd m) &= g \circ f|_{e_v M}(e_v \rhd m) \\
&= e_v \rhd (g \circ f)(m) \\
&= g(e_v \rhd f(m)) \\
&= g(\mathcal{R}(M)(f)_v(e_v \rhd m)) \\
&= (\mathcal{R}(M)(g)_v \circ \mathcal{R}(M)(f)_v)(e_v \rhd m)
\end{aligned}
$$

and one may readily verify that $\mathcal{R}(M)(\mathrm{id}_M)_v = \mathrm{id}_{e_v M}$. $\square$

Remark 5.3.8. The preceding proof is not completely straightforward to parse: we recommend drawing appropriate commutative diagrams in order to trace through what is being computed.

Proposition 5.3.9. *The functors $\mathcal{M}$ and $\mathcal{R}$ are inverse isomorphisms of categories.*

Proof. On objects, we have $e_v \left(\bigoplus_w Fw \right) = Fv$ and $\bigoplus_v \mathcal{R}(M)_w = \bigoplus_v e_v M = M$. On morphisms, we also have the equalities $\mathcal{R}(\mathcal{M}F)(\mathcal{M}(\alpha))_v = \alpha_v$ and $\bigoplus_v f|_{e_v M} = f$. Hence $\mathcal{R} \circ \mathcal{M} = \mathrm{id}_{\mathrm{Rep}(\mathcal{Q})}$ and $\mathcal{M} \circ \mathcal{R} = \mathrm{id}_{\mathbb{K}[\mathcal{Q}]\text{-Mod}}$ as functors. $\square$

That is,

> Linear representations of $\mathcal{Q}$ correspond to $\mathbb{K}[\mathcal{Q}]$-modules.

Now that we have established that we can study the representation theory of quivers via that of the path algebra, we can ask which modules have the properties we studied in Chapter 4. However, it will be convenient to use the dictionary above in both directions, that is, where this makes the example clearer, we will often write down a representation as well as, or even instead of, the module.

5.3.2   Simple, projective and injective modules

We will start with simple modules. Since 1-dimensional vector spaces have no proper non-zero subspaces, any 1-dimensional module must be simple. In terms of representations, we should consider representations whose dimension function takes value 1 at some vertex v and 0 everywhere else.

Recall that $\mathcal{Q}$ is assumed to be a finite quiver throughout.

Definition 5.3.10. Let $\mathcal{Q}$ be a quiver and $\mathbb{K}$ a field. For $v \in \mathrm{Vert}(\mathcal{Q})$, define $S_v \colon \mathcal{P}(\mathcal{Q}) \to \mathbb{K}\text{-Mod}$ by

$$
S_v w = \begin{cases} \mathbb{K} & \text{if } v = w \\ 0 & \text{otherwise} \end{cases}
$$

and

$$S_v p = \begin{cases} \mathrm{id}_{\mathbb{K}} & \text{if } p = e_v \\ 0 & \text{otherwise} \end{cases}$$

We abuse notation (suppressing the functor $\mathcal{M}$) and write S_v also for the module $\mathcal{M}(S_v)$ having underlying vector space $\mathbb{K}$ and action $e_v \rhd x = x$ and $p \rhd x = (S_v p)(x) = 0$ for $p \neq e_v$, for all $x \in \mathbb{K}$.

Example 5.3.11. Let $\mathcal{Q} = A_3$ be the quiver $1 \longrightarrow 2 \longrightarrow 3$. Then S_1, S_2 and S_3 are the representations

$$S_1 \qquad \mathbb{K} \xrightarrow{\;0\;} 0 \xrightarrow{\;0\;} 0$$

$$S_2 \qquad 0 \xrightarrow{\;0\;} \mathbb{K} \xrightarrow{\;0\;} 0$$

$$S_3 \qquad 0 \xrightarrow{\;0\;} 0 \xrightarrow{\;0\;} \mathbb{K}$$

with dimension vectors $(1,0,0)$, $(0,1,0)$ and $(0,0,1)$ respectively; recall that we don't draw the identity maps at each vertex but it is important to remember that they are there. Note too that S_2 appeared as E in Example 3.3.4.

Rather than construct representations "by hand" as we did in Example 3.3.4, SageMath knows how to make simple quiver representations and hence how to test if a representation is simple.

```
1  Q = DiGraph({1:{2:['e12']},2:{3:['e23']}})
2  PQ = Q.path_semigroup()
3  E = PQ.representation(QQ,{1: QQ^0, 2: QQ^1, 3: QQ
      ^0},{(1,2,'e12'): [], (2,3,'e23'): []})
4  F = PQ.representation(QQ,{1: QQ^0, 2: QQ^1, 3: QQ
      ^1},{(1,2,'e12'): [], (2,3,'e23'): [1]})
5  F.is_simple()
6  S2 = PQ.S(QQ,2)
7  S2 == E
```

Via line 5, SageMath confirms that F is not simple. In line 6, we ask SageMath to construct the simple PQ-module (i.e. representation of Q) over the field QQ (i.e. $\mathbb{Q}$) at vertex 2. In line 7, we check that this is the same as E.

The associated modules S_i, $i = 1, 2, 3$, all have underlying vector space $\mathbb{K}$ but they are not isomorphic modules as the actions differ: S_1 has $e_1 \rhd_1 x = x$

and $e_2 \rhd_1 x = 0$ for any x, whereas S_2 has $e_1 \rhd_2 x = 0$ and $e_2 \rhd_2 x = x$.

This is why we often prefer quiver representations to modules in examples: when we talk about modules $(M, \rhd)$ we tend to suppress the action and just say "M" but since even the dimension vector is not enough to specify the representation, it is certainly not the case that knowing the underlying vector space for a module is anywhere near enough to distinguish them. If we ignored the actions, we would—erroneously—think that S_1, S_2 and S_3 were the same.

These modules are indeed simple: S_v is 1-dimensional and so must be simple, as above. However, not every simple module must be of this form. If $Q = L_1$ is the quiver with one vertex v and a loop α at that vertex, then for each $\lambda \in \mathbb{K}$ we have a simple representation

$$S(\lambda) \qquad \mathbb{K} \;\circlearrowleft\; \lambda I$$

and $S(\lambda) \cong S(\mu)$ if and only if $\lambda \neq \mu$. Note that $S(0) = S_v$.

We now examine the projective and injective modules for path algebras. As we do so, you might wish to compare with Example 3.3.4.

Proposition 5.3.12. *For each $v \in \mathrm{Vert}(Q)$, $P_v \overset{\mathrm{def}}{=} \mathbb{K}[Q]e_v$ is a projective module.*

Proof. Since Q is a finite quiver, as in Lemma 5.3.1, we can write $1_{\mathbb{K}[Q]} = \sum_v e_v$ where the e_v are orthogonal idempotents. Then it is a general fact that $_{\mathbb{K}[Q]}\mathbb{K}[Q] = \bigoplus_v (_{\mathbb{K}[Q]}\mathbb{K}[Q])e_v$, as left modules.

Now by Proposition 4.5.4, $_{\mathbb{K}[Q]}\mathbb{K}[Q]e_v$ is projective, being a direct summand of a free module. $\qquad \square$

Lemma 5.3.13. *The module P_v has a quotient isomorphic to S_v.*

Proof. We construct a surjective module homomorphism onto S_v as follows. We have $P_v = \mathbb{K}[Q]e_v$ so that every basis element of P_v has the form pe_v for some path p; if $t(p) \neq v$ then $pe_v = 0$ so we may assume that $t(p) = v$. Recall that S_v has underlying vector space $\mathbb{K}$, so that $\{1_\mathbb{K}\}$ is a basis.

Define a linear map on the basis of P_v by $\pi_v \colon P_v \to S_v$ by

$$\pi_v(pe_v) = \begin{cases} 1_\mathbb{K} & \text{if } p = e_v \\ 0 & \text{otherwise} \end{cases}$$

and extend linearly. This is a module homomorphism: let q be a path in Q. Then $q \rhd \pi_v(pe_v) = 0$ unless $p = q = e_v$ and $e_v \rhd \pi_v(e_v e_v) = e_v \rhd 1_\mathbb{K} = 1_\mathbb{K}$. On the other hand, $\pi_v(q \rhd pe_v) = \pi_v(qpe_v) = 0$ unless $qp = e_v$, in which case we must have $q = p = e_v$, and $\pi_v(e_v \rhd e_v e_v) = \pi_v(e_v) = 1_\mathbb{K}$ in this case.

It is clear that π_v is surjective and so $P_v / \mathrm{Ker}\, \pi_v \cong S_v$. $\qquad \square$

Note that Ker π_v has a basis consisting of all paths in $\mathcal{Q}$ starting at v and of length at least 1.

Unfortunately, it is not so straightforward to identify the injective modules. We will introduce two natural collections of candidates but if $\mathcal{Q}$ is not acyclic they might fail to be injective or might fail to yield all the injective modules.

Let V be a $\mathbb{K}$-vector space. Denote by V^* the $\mathbb{K}$-dual of V, that is, the vector space $V^* = \mathrm{Hom}_{\mathbf{Vect}_\mathbb{K}}(V, \mathbb{K})$ of linear functionals on V, with operations being pointwise addition and scalar multiplication.

Problem 33. Let $\mathbb{K}$ be a field and A a $\mathbb{K}$-algebra. If $M = (M, \triangleright)$ is a left A-module, then M^* is naturally a right A-module, via $(f \triangleleft a)(m) \stackrel{\text{def}}{=} f(a \triangleright m)$. If $(M, \triangleleft)$ is a right A-module, then M^* is naturally a left A-module.

(Be warned, however, that if M is not finite-dimensional, it is possible that $M \not\cong M^{**}$; the latter can be strictly larger.)

Definition 5.3.14. For each $v \in \mathrm{Vert}(\mathcal{Q})$, define $_vP = e_v\mathbb{K}[\mathcal{Q}]$ to be the *right* $\mathbb{K}[\mathcal{Q}]$-module with basis all paths in $\mathcal{Q}$ ending in v with action $e_vp\triangleleft q = e_vpq$.

For the same reason as for P_v, $_vP$ is a projective right $\mathbb{K}[\mathcal{Q}]$-module, being a direct summand of the free module $\mathbb{K}[\mathcal{Q}]$.

Proposition 5.3.15. *Let $v \in \mathrm{Vert}(\mathcal{Q})$ be such that $_vP$ is finite-dimensional, that is, only finitely many paths in $\mathcal{Q}$ end in v. Then $(_vP)^*$ is an injective left $\mathbb{K}[\mathcal{Q}]$-module, with respect to the induced module structure.*

Proof. By the above exercise, it suffices to show that $(_vP)^*$ is injective. We will use characterization (iii) from Proposition 4.5.5. Let $f: (_vP)^* \hookrightarrow M$ be a monomorphism.

Since $_vP$ is finite-dimensional, there is an isomorphism $\iota: (_vP)^{**} \to {}_vP$, whose inverse is the map $\iota^{-1}: {}_vP \to (_vP)^{**}$ given by $\iota^{-1}(x): (_vP)^* \to \mathbb{K}$, $\iota^{-1}(x)(\varphi) = \varphi(x)$. Then $f^*: M^* \twoheadrightarrow (_vP)^{**}$ is a surjection onto $(_vP)^{**}$ and thus $g: M^* \twoheadrightarrow {}_vP$, $g = \iota \circ f^*$ is a surjection onto $_vP$.

As $_vP$ is projective, the right module analogue of Proposition 4.5.4(iii) yields a section $s: {}_vP \hookrightarrow M^*$ such that $g \circ s = \mathrm{id}_{_vP}$. Then $s^*: M^{**} \twoheadrightarrow (_vP)^*$ satisfies $s^* \circ g^* = s^* \circ f^{**} \circ \iota^* = \mathrm{id}_{(_vP)^*}$.

Although not necessarily an isomorphism, there is an injective module homomorphism $\iota_M: M \hookrightarrow M^{**}$, given by $m \mapsto \iota_M(m): M^* \to \mathbb{K}$ with $\iota_M(m)(\lambda) = \lambda(m)$ for all $\lambda \in M^*$.

Now defining $r = s^* \circ \iota_M$ we have a surjective module homomorphism

$r\colon M \twoheadrightarrow (_vP)^*$ and for $\varphi \in (_vP)^*$ and $x \in {_vP}$ we have

$$
\begin{aligned}
(r \circ f)(\varphi)(x) &= ((s^* \circ \iota_M)(f(\varphi)))(x) \\
&= s^*(\iota_M(f(\varphi)))(x) \\
&= \iota_M(f(\varphi))(s(x)) \\
&= s(x)(f(\varphi)) \\
&= f^*(s(x))(\varphi) \\
&= \iota^{-1}(x)(\varphi) \\
&= \varphi(x)
\end{aligned}
$$

so that $r \circ f = \mathrm{id}_{(_vP)^*}$ as desired. $\qquad\qquad\qquad\qquad\qquad\square$

Remark 5.3.16. Examining the proof, we see that we have used no information about $_vP$ other than that it is finite-dimensional and projective. Therefore, we have in fact proved that the dual of a finite-dimensional projective right module is an injective left module, in general. (Again, we suggest drawing appropriate commutative diagrams to assist in understanding the proof.)

SageMath can construct the projective and injective modules (under suitable assumptions); this code should be used in conjunction with that above in Example 5.3.11.

```
8   S1 = PQ.S(QQ,1); S3 = PQ.S(QQ,3)
9   P1 = PQ.P(QQ,1); P2 = PQ.P(QQ,2); P3 = PQ.P(QQ,3)
10  I1 = PQ.I(QQ,1); I2 = PQ.I(QQ,2); I3 = PQ.I(QQ,3)
11  P2.quotient(S3)
12  pi = P2.hom({1: [], 2:[1], 3:[]},S2)
13  pi.is_injective()
14  pi.is_surjective()
15  pi.kernel()
16  pi.image()
```

Think about how this code relates to Lemma 5.3.13 and, using the documentation (https://doc.sagemath.org/html/en/reference/quivers/sage/quivers/representation.html), experiment with other relationships among the simples, projectives and injectives. This is good preparation for understanding Examples 6.2.6(b).

Let p be a path in Q. We say that a path q *divides* p and write $q \mid p$ if there exists a path r such that $p = r \circ q$; such a path r is unique if it exists. That is, $q \mid p$ if the path p begins with the path q. If $q \mid p$, define pq^{-1} to be the (unique) path such that $p = (pq^{-1}) \circ q$. Note that $h(pq^{-1}) = h(p)$.

Definition 5.3.17. Let $v \in \mathrm{Vert}(\mathcal{Q})$. Let I_v be the $\mathbb{K}[\mathcal{Q}]$-module with underlying vector space $\bigoplus_{u \in \mathrm{Vert}(\mathcal{Q})} \mathbb{K}[\mathcal{P}(u,v)] = e_v \mathbb{K}[\mathcal{Q}]$ (i.e. the subspace of $\mathbb{K}[\mathcal{Q}]$ spanned by all paths ending in v) and action

$$q \triangleright e_v p = \begin{cases} e_v p q^{-1} & \text{if } q \mid p \\ 0 & \text{otherwise} \end{cases}$$

To see that this is a $\mathbb{K}[\mathcal{Q}]$-module, consider $r \triangleright (q \triangleright e_v p)$. If $q \nmid p$,

$$r \triangleright (q \triangleright e_v p) = r \triangleright 0 = 0.$$

Furthermore, if $q \nmid p$ then $rq \nmid p$ also, so

$$rq \triangleright e_v p = 0 = r \triangleright (q \triangleright e_v p).$$

Now assume $q \mid p$. Then

$$r \triangleright (q \triangleright e_v p) = r \triangleright e_v p q^{-1}.$$

If $r \nmid pq^{-1}$, $r \triangleright e_v p q^{-1} = 0$. Furthermore, if $r \nmid pq^{-1}$ then $rq \nmid p$ and

$$rq \triangleright e_v p q^{-1} = 0 = r \triangleright (q \triangleright e_v p).$$

This leaves the case $q \mid p$ and $r \mid pq^{-1}$. So there exists $p(rq)^{-1}$ such that $pq^{-1} = p(rq)^{-1} \circ r$ and hence $p = pq^{-1} \circ q = p(rq)^{-1} \circ r \circ q$. That is, $rq \mid p$ also. Hence

$$r \triangleright (q \triangleright e_v p) = r \triangleright e_v p q^{-1} = e_v p (rq)^{-1}$$

and

$$rq \triangleright e_v p = e_v p (rq)^{-1}$$

also, so that $\triangleright$ is indeed a module structure.

Lemma 5.3.18. *The module I_v has a submodule isomorphic to S_v.*

Proof. Let S be the subspace of $I_v = e_v \mathbb{K}[\mathcal{Q}]$ spanned by $e_v e_v = e_v$. Then for all $p \in \mathbb{K}[\mathcal{Q}]$, we claim that $p \mid e_v$ if and only if $p = e_v$.

If $p = e_v$ then $e_v = e_v \circ e_v$ so $e_v \mid e_v$; then $e_v e_v^{-1} = e_v$.

If $p \neq e_v$, either $p = e_w$ for $w \neq v$ or p has length at least 1. If $p = e_w$ then any r such that $e_v = r \circ e_w$ would have to have length 0, so $r = e_x$ for some x. However $e_v = e_x \circ e_w$ can hold if and only if $v = x = w$, contradicting $v \neq w$. If p has length at least 1, there is no path r such that $e_v = r \circ p$, by considering the lengths of both sides. In either situation, we see that $p \nmid e_v$.

Hence $p \triangleright e_v = 0$ for all $p \neq e_v$ and $e_v \triangleright e_v = e_v e_v = e_v$. This shows that S is isomorphic to S_v. $\square$

If our quiver is acyclic (that is, has no oriented cycles) we may make the following stronger statements.

Proposition 5.3.19. *Let Q be an acyclic quiver.*

  (i) *The collection $\mathbb{S} = \{S_v \mid v \in \mathrm{Vert}(Q)\}$ is a complete set of pairwise non-isomorphic simple $\mathbb{K}[Q]$-modules.*

     That is, every simple $\mathbb{K}[Q]$-module is isomorphic to a member of $\mathbb{S}$ and no two distinct members of $\mathbb{S}$ are isomorphic.

  (ii) *The collection $\mathbb{P} = \{P_v \mid v \in \mathrm{Vert}(Q)\}$ is a complete set of non-isomorphic indecomposable projective modules.*

  (iii) *We have $I_v \cong (_vP)^*$ for each v and the collection $\mathbb{I} = \{I_v \mid v \in \mathrm{Vert}(Q)\}$ is a complete set of non-isomorphic indecomposable injective modules.*

  (iv) *We have $P_v \cong S_v$ if and only if v is a sink (i.e. there are no arrows α with $t(\alpha) = v$).*

  (v) *We have $I_v \cong S_v$ if and only if v is a source (i.e. there are no arrows α with $h(\alpha) = v$).*

Proof.

  (i) We first show that if M is any non-zero $\mathbb{K}[Q]$-module, M has a submodule isomorphic to S_v for some v. Since Q is acyclic, there exists a vertex $v \in \mathrm{Vert}(Q)$ such that $e_v M \neq 0$ and $e_w M = 0$ for any arrow $\alpha \in Q(v, w)$ (exercise). Let $m \in e_v M$, $m \neq 0$ and define $f \colon S_v \to M$ to be the injective linear map defined by $f(1_{\mathbb{K}}) = m$.

    Now for any arrows $\alpha \in Q(u, v)$ and $\beta \in Q(v, w)$, the diagram

$$
\begin{array}{ccccc}
0 & \xrightarrow{\ 0\ } & e_v S_v & \xrightarrow{\ 0\ } & 0 \\
{\scriptstyle 0}\downarrow & & {\scriptstyle \mathcal{R}(M)(f)}\downarrow & & \downarrow{\scriptstyle 0} \\
e_u M & \longrightarrow & e_v M & \longrightarrow & e_w M = 0 \\
& & \mathcal{R}(M)(\beta)=0 & &
\end{array}
$$

commutes; note that we use the particular choice of v to know that the bottom right-most space is zero. As a result, f is a module homomorphism and being injective, this proves that M has a submodule isomorphic to S_v. As S_v is simple, either $M \cong S_v$ or M is not simple.

    That the S_v are pairwise non-isomorphic follows as in the discussion of Example 5.3.11.

  (ii) Let us show that P_v is indecomposable. Firstly, since Q is acyclic, $e_v P_v$ is 1-dimensional, spanned by the only path from v to itself in Q, namely e_v.

    Now if $P_v = M \oplus N$ with M, N non-zero, we may assume (without loss of generality, by interchanging the roles of M and N) that $e_v P_v = e_v M$

and $e_v N = 0$. Since N is non-zero, there exists w such that $e_w N \neq 0$. Now $e_w P_v$ has as basis all paths from v to w; let p be such a path.

Then $\mathcal{R}(P_v)(p)\colon e_v P_v = e_v M \oplus 0 \to e_w P_v = e_w M \oplus e_w N$ sends $e_v \oplus 0$ to $\mathcal{R}(P_v)(p)(e_v \oplus 0) = p \triangleright (e_v \oplus 0) = p \oplus 0$. So every basis element of $e_w P_v$ in fact lies in $e_w M$, but this contradicts $e_w N \neq 0$. We therefore conclude that P_v has no non-trivial decomposition into non-zero M and N.

Since ${}_{\mathbb{K}[\mathcal{Q}]}\mathbb{K}[\mathcal{Q}] = \bigoplus_v P_v$ and the P_v are indecomposable, any direct summand of a free module—that is, any projective module—is isomorphic to a direct sum of the P_v.

(iii) Since $\mathcal{Q}$ is acyclic, the right modules ${}_v P$ are finite-dimensional for all $v \in \mathrm{Vert}(\mathcal{Q})$ and hence Proposition 5.3.15 tells us that the $({}_v P)^*$ are injective for all v. We leave it as an exercise to show (i) that I_v is isomorphic to $({}_v P)^*$ *as modules*, noting that I_v and ${}_v P$ have the same underlying vector space so $({}_v P)^* = I_v^* \cong I_v$ as vector spaces (where the isomorphism here is the *un*natural isomorphism of vector spaces induced by the dual basis construction); and (ii) argue similarly to the previous part to show that these are a complete set of indecomposable injectives.

(iv) From the proof of Lemma 5.3.13, we know that there is a surjective homomorphism $\pi_v \colon P_v \twoheadrightarrow S_v$ with kernel all paths in $\mathcal{Q}$ starting at v and of length at least 1. If v is a sink, since $\mathcal{Q}$ is acyclic, there are no such paths, and so π_v is an isomorphism of P_v with S_v.

Conversely, if $P_v \cong S_v$, acyclicity implies that the only path in $\mathcal{Q}$ starting at v is e_v, i.e. v is a sink.

(v) The proof of Lemma 5.3.18 tells us that there is an injective homomorphism $i_v \colon S_v \hookrightarrow I_v$. If v is a source, since $\mathcal{Q}$ is acyclic, there is a unique path in $\mathcal{Q}$ ending in v, namely e_v. So I_v is 1-dimensional and i_v is an isomorphism of I_v with S_v.

Conversely, if $I_v \cong S_v$, acyclicity implies that the only path in $\mathcal{Q}$ ending at v is e_v, i.e. v is a source. $\square$

We emphasise that if $\mathcal{Q}$ is acyclic, then ${}_{\mathbb{K}[\mathcal{Q}]}\mathbb{K}[\mathcal{Q}] \cong \bigoplus_{v \in \mathrm{Vert}(\mathcal{Q})} P_v$. Again, we suggest looking back to Examples 3.3.4 and 5.3.11 and perhaps even extending what was there further.

5.3.3 Semisimplicity and representation type

Given the importance we laid on semisimplicity for groups, it is natural for us to next consider when the path algebra is semisimple. The situation is a little more complicated than the groups case; however, we can give a complete classification for the acyclic case.

Proposition 5.3.20. *Let Q be an acyclic quiver. Then $\mathbb{K}[Q]$ is semisimple if and only if Q has no arrows.*

We defer the proof temporarily, to give two additional definitions and a theorem, from which we will derive this result.

Definition 5.3.21. Let A be a $\mathbb{K}$-algebra over a field $\mathbb{K}$. Define the *Jacobson radical* $\mathcal{J}(A)$ of A to be the intersection of all maximal left ideals of A (equivalently, the intersection of all maximal A-submodules of $_A A$).

Definition 5.3.22. Let A be a $\mathbb{K}$-algebra over a field and let M be an A-module. Define the *annihilator* $\mathrm{Ann}_A(M)$ to be the set

$$\mathrm{Ann}_A(M) = \{a \in A \mid a \triangleright m = 0 \text{ for all } m \in M\}.$$

The annihilator of a module M is in fact a two-sided ideal of A (exercise).

Theorem 5.3.23 (cf. [EH, Theorem 4.23]). *Let A be a $\mathbb{K}$-algebra over a field $\mathbb{K}$ and assume that $_A A$ has finite length. Then*

(i) *$\mathcal{J}(A)$ is the intersection of finitely many maximal left ideals of A;*

(ii) *$\mathcal{J}(A) = \bigcap_{S \text{ simple}} \mathrm{Ann}_A(S)$;*

(iii) *$\mathcal{J}(A)$ is a two-sided ideal of A;*

(iv) *$A/\mathcal{J}(A)$ is a semisimple $\mathbb{K}$-algebra;*

(v) *A is semisimple if and only if $\mathcal{J}(A) = 0$.*

Proof. Omitted—see the reference given above. □

Proposition 5.3.24. *Let Q be an acyclic quiver. Then the Jacobson radical $\mathcal{J}(\mathbb{K}[Q])$ of $\mathbb{K}[Q]$ is the submodule of $\mathbb{K}[Q]$ spanned by all paths in Q of strictly positive length.*

Proof. For $v \in \mathrm{Vert}(Q)$, define $\mathcal{J}_v$ to be the left ideal of $\mathbb{K}[Q]$ spanned by all paths in Q starting at v and having strictly positive length. Then $S_v \cong P_v/\mathcal{J}_v$ and one can check that

$$\mathrm{Ann}_{\mathbb{K}[Q]}(S_v) = \mathcal{J}_v \oplus \bigoplus_{w \neq v} P_w.$$

Taking the intersection of all of these, we deduce from Theorem 5.3.23(ii) that $\mathcal{J}(\mathbb{K}[Q]) = \bigoplus_{v \in \mathrm{Vert}(Q)} \mathcal{J}_v$, the latter being the span of *all* paths in Q of strictly positive length. □

Now we may prove Proposition 5.3.20:

Proof of Proposition 5.3.20: By Theorem 5.3.23, $\mathbb{K}[\mathcal{Q}]$ is semisimple if and only if $\mathcal{J}(\mathbb{K}[\mathcal{Q}]) = 0$. By the characterization in Proposition 5.3.24, the Jacobson radical being zero implies that there are no paths of strictly positive length in $\mathcal{Q}$. So if $\mathbb{K}[\mathcal{Q}]$ is semisimple, $\mathcal{Q}$ must have no arrows.

For the converse, if $\mathcal{Q}$ has no arrows, $\mathbb{K}[\mathcal{Q}] \cong \prod_{v \in \mathrm{Vert}(\mathcal{Q})} \mathbb{K}$ as algebras so is clearly semisimple.   □

Remark 5.3.25. Note that if $\mathcal{Q}$ is not assumed to be acyclic, we need to take care. It remains true that if $\mathcal{Q}$ has no arrows, then $\mathbb{K}[\mathcal{Q}]$ is semisimple.

However, for the quiver L_1 with one vertex and one loop

$$L_1 \qquad 1 \;\circlearrowleft\; \alpha$$

we saw earlier that $\mathbb{K}[L_1]$ has infinitely many non-isomorphic simple modules. If it were semisimple, this would contradict Proposition 4.5.8. Indeed, $\mathbb{K}[L_1] \cong \mathbb{K}[x]$, the polynomial algebra in one variable, which fails to be semisimple in several different ways.

This algebra does have zero Jacobson radical, however; clearly in this case, zero Jacobson radical is not implying no arrows. Note that although the claim that the Jacobson radical is zero remains true whether the field is finite or infinite, it is significantly harder to prove in the finite case.

The issue here is that characterizing semisimplicity for infinite-dimensional algebras is not straightforward. In the case of groups, Maschke's theorem (5.2.13) applies to *finite* groups, for which the group algebra is finite-dimensional; we did not have a more general claim there either.

Remark 5.3.26. Other than the penultimate paragraph of the previous remark, we have not needed to consider properties of the base field, unlike in Maschke's theorem.

It is clear from the above results that if we were to say, as one might be tempted to by analogy with the groups situation, "let us only worry about semisimple path algebras", there would not be much of interest left. So, for quivers, we should not ask for something as strong as being semisimple. Perhaps we could still say a lot by looking at decompositions into indecomposable (but not necessarily simple) modules.

Recall that we have the Krull–Remark–Schmidt theorem: if M is a finite length module over an R-algebra A, it has an essentially unique decomposition into a finite number of indecomposable submodules. We could hope that A has a collection of indecomposable modules we can classify up to isomorphism, for then we would understand all finite length modules.

To avoid having to handle more complicated definitions and other subtleties that arise for general R-algebras, we will now restrict again to $R = \mathbb{K}$ being a field. Furthermore, we will use finite versus infinite dimension (rather than length) as our discriminating property, in line with usual practice in the study of finite-dimensional algebras, where there is no difference, noting that this means we will often have acyclic quivers in mind.

Definition 5.3.27. Let A be a $\mathbb{K}$-algebra over a field $\mathbb{K}$. We say that A has *finite representation type* if A has finitely many finite-dimensional indecomposable A-modules, up to isomorphism. Otherwise, A is said to have *infinite representation type*.

It is common for people to say "representation finite" for "finite representation type" and similarly "representation infinite" for "infinite representation type".

We give three examples from the representation theory of quivers and one theorem from the representation theory of finite groups.

Examples 5.3.28.

(a) Consider the quiver A_2 with two vertices and one arrow, $1 \xrightarrow{\ \alpha\ } 2$.

The quiver A_2 is acyclic and we can use what we learned about this case above in Proposition 5.3.19 to describe the indecomposable modules. Firstly, there are exactly two 1-dimensional simple modules, S_1 and S_2. Since 1 is a source, $S_1 \cong I_1$ is injective; since 2 is a sink, $S_2 \cong P_2$ is projective.

The other projective module is P_1, a 2-dimensional indecomposable module with basis all paths starting at 1, i.e. $\{e_1, \alpha\}$. The other injective module is I_2, a 2-dimensional indecomposable module with basis all paths ending at 2, i.e. $\{\alpha, e_2\}$. Note that P_1 has a quotient isomorphic to S_1, arising from the map sending e_1 to e_1 and α to 0; similarly I_2 has a submodule isomorphic to S_2, this being the submodule spanned by e_2.

We have that $_{\mathbb{K}[A_2]}\mathbb{K}[A_2] \cong P_1 \oplus P_2$ and we see this explicitly by noting that a basis for $_{\mathbb{K}[A_2]}\mathbb{K}[A_2]$ is $\{e_1, \alpha, e_2\}$, which is the union of the bases of P_1 and $P_2 = S_2$.

We claim that $P_1 \cong I_2$. Let $f \colon P_1 \to I_2$ be the linear map defined by $f(e_1) = \alpha$ and $f(\alpha) = e_2$. Then

$$f(e_1 \rhd_1 e_1) = f(e_1) = \alpha = e_1 \rhd_2 \alpha = e_1 \rhd_2 f(e_1)$$
$$f(\alpha \rhd_1 e_1) = f(\alpha) = e_2 = \alpha \rhd_2 \alpha = \alpha \rhd_2 f(e_1)$$
$$f(e_2 \rhd_1 e_1) = f(0) = 0 = e_2 \rhd_2 \alpha = e_2 \rhd_2 f(e_1)$$

and similarly for α; here, $\rhd_1$ refers to the action on P_1 and $\rhd_2$ that on I_2. Since f is by definition bijective on the two bases, it is a linear and hence module isomorphism.

This gives us three finite-dimensional indecomposable modules for $\mathbb{K}[A_2]$, $S_1 \cong I_1$, $I_2 \cong P_1$ and $S_2 \cong P_2$, as well as a short exact sequence

$$0 \longrightarrow S_2 \lhook\joinrel\longrightarrow I_2 \longrightarrow\!\!\!\!\!\rightarrow S_1 \longrightarrow 0$$

This sequence is not split: I_2 is a non-trivial extension of S_2 by S_1. However it is a particularly nice extension; see Section 6.2 for more on this.

We might jump to thinking that therefore A_2 has finite representation type. However, we have not quite proved this yet: there could be other finite-dimensional indecomposable modules that are not of the forms we know.

The standard way to "know" that other such modules do not exist is to invoke *Gabriel's theorem*; see Section 5.3.4. But for A_2 specifically, this is a direct consequence of the existence of Smith normal form (Theorem 5.1.9)! So, A_2 is of finite representation type.

(b) Consider the *Kronecker quiver* K_2 $1 \overset{\longrightarrow}{\longrightarrow} 2$.

This quiver is also acyclic but for each $\lambda \in \mathbb{K}$, there is an indecomposable representation C_λ given by

$$ C_\lambda = \qquad \mathbb{K} \underset{\lambda \mathrm{id}_{\mathbb{K}}}{\overset{\mathrm{id}_{\mathbb{K}}}{\rightrightarrows}} \mathbb{K} $$

and $C_\lambda \cong C_\mu$ if and only if $\lambda = \mu$. Therefore if $\mathbb{K}$ is an infinite field, we deduce that the path algebra $\mathbb{K}[K_2]$ has infinite representation type. Note that with a little more ingenuity, the result can be shown to remain true for finite fields also; replace $\mathbb{K}$ by $\mathbb{K}^n$, 1 by the identity matrix and λ by the Jordan block $J_n(1)$ (see [EH, Example 9.30] for details).

(c) We have seen that the quiver L_1 with one vertex and one loop has (at least) as many non-isomorphic simple modules of dimension 1 as there are elements of the field $\mathbb{K}$, so if $\mathbb{K}$ is an infinite field, the associated path algebra has infinite representation type.

Theorem 5.3.29 ([EH, Theorem 8.17]). *Let $\mathbb{K}$ be a field and G a finite group. Then $\mathbb{K}[G]$ has finite representation type if and only if $\mathbb{K}$ has characteristic zero or $\mathbb{K}$ has characteristic $p > 0$ and G has a cyclic Sylow p-subgroup.*

The statement of this theorem uses a small amount of advanced group theory, which may be found in most textbooks on the subject (e.g. [AB], [ST]). You might also like to reflect on how it meshes with Maschke's theorem (5.2.13).

5.3.4   Gabriel's theorem

Gabriel's theorem does for representation type of path algebras what Maschke's theorem does for semisimplicity of group algebras, namely it provides a concrete way to verify whether the path algebra of an acyclic quiver is of finite representation type.

We will not prove this theorem: doing so is one of the main aims of [EH] and requires substantially more space than we have here. Instead, we will define the ingredients of the classification and state the theorem.

Definition 5.3.30. Let Q be a quiver. Then Q has an underlying (undirected) graph $\Gamma(Q)$ with vertex set $\mathrm{Vert}(Q)$ and (undirected) edges

$$\{\{h(\alpha), t(\alpha)\} \mid \alpha \in Q(v, w) \text{ for some } v, w \in \mathrm{Vert}(Q)\}$$

That is, the edges of $\Gamma(Q)$ are given by forgetting the orientation on each arrow.

We say that a quiver Q is an orientation of an undirected graph Γ' if $\Gamma(Q) = \Gamma'$, that is, Q is obtained by choosing an orientation for each edge of Γ'.

The following definition describes an extremely important class of undirected graphs. We could not possibly do justice in a few words to why this class is important; entire books are written about this topic, and we encourage you to go read them (after finishing this one, of course).

Definition 5.3.31. The following collection of undirected graphs are the *simply-laced Dynkin diagrams of finite type*:

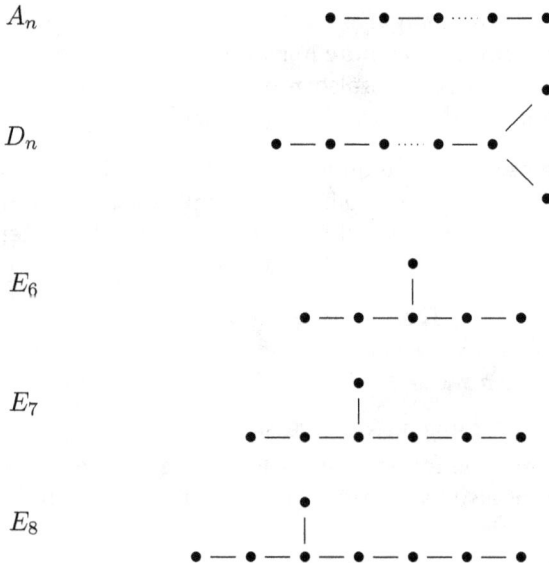

Here the subscript on the name denotes the number of vertices of the graph.

Theorem 5.3.32 (Gabriel's theorem). *Let Q be an acyclic quiver. Then $\mathbb{K}[Q]$ is of finite representation type if and only if Q is the disjoint union of finitely many quivers, each of which is an orientation of a simply-laced Dynkin diagram of finite type.*

The remarkable thing about Gabriel's theorem (even more remarkable than just taking its statement at face value) is that its proof proves even more:

Theorem 5.3.33. *Let Q be an acyclic quiver such that $\mathbb{K}[Q]$ is of finite representation type. Then the indecomposable representations of $\mathbb{K}[Q]$ are in bijection with the set of positive roots of the root system associated to Q, under a bijection sending a module to its dimension vector.*

Here, a *root system* is a particular configuration of vectors in a Euclidean space that is naturally associated to a Lie algebra. We will not give details here but refer the interested reader to [FH] and similar books.

Instead, we just note that A_2 has three positive roots in its associated root system, α_2, $\alpha_2 + \alpha_1$ and α_1. We have written them in a somewhat esoteric order because they then line up with the three indecomposable modules we know, S_2, I_2 (which is a non-split extension of S_2 and S_1) and S_1.

Thus Gabriel's theorem identifies a deep and fruitful link between the representation theory of quivers and the theory of Lie algebras.

5.E Exercises

Exercise 5.1. Prove Lemma 5.1.13.

Exercise 5.2. Using Example 5.1.12 as a template, classify the Abelian groups of order n up to isomorphism for at least three values of $n \geq 100$ that you choose.

Exercise 5.3 (cf. Problem 28). Create some examples of matrices with different Smith normal, rational canonical and Jordan normal forms.

Exercise 5.4.

(a) Let $\mathbb{C}[C_5]$ be the group algebra over the complex numbers of the finite cyclic group of order 5, $C_5 = \langle a \mid a^5 = e \rangle$.

    (a) Write down a basis for $\mathbb{C}[C_5]$ and hence write down an expression for a typical element of $\mathbb{C}[C_5]$.

    (b) Construct a multiplication table for this basis. (That is, if your basis is $\mathcal{B} = \{b_1, b_2, \dots\}$, write down the values of $b_i b_j$ for all $b_i, b_j \in \mathcal{B}$.)

(b) Let $A = \mathbb{C}[x]/I$ be the algebra given by the quotient of the polynomial algebra in one variable over $\mathbb{C}$, $\mathbb{C}[x]$, by the ideal $I = \langle x^5 - 1 \rangle$ generated by $x^5 - 1$. Using the fact that $\mathcal{B}' = \{1, x, x^2, x^3, \dots\}$ is a basis for $\mathbb{C}[x]$ and that hence the set $\{1 + I, x + I, x^2 + I, x^3 + I, \dots\}$ spans A, find a basis for A.

    [Hint: what can you say about $x^5 + I$ and $1 + I$?]

(c) Hence construct an isomorphism of algebras $f \colon \mathbb{C}[C_5] \to A$.

In relation to the previous exercise, you might find the following helpful.

```
1   R.<x> = PolynomialRing(CC)
2   S = R.quotient(x^5 - 1, 'y'); y=S.gen()
3   for i in range(6):
4           print(y^i)
```

Exercise 5.5. Let $G = \langle a, b \mid a^2 = b^2 = (ab)^2 = e \rangle \cong C_2 \times C_2$ and M a $\mathbb{C}$-vector space of dimension 2 with basis $\mathcal{B}_M = \{m_1, m_2\}$. Which of the following actions of the generators a, b makes M into a $\mathbb{C}[G]$-module?

(a) $a \triangleright m_1 = m_2,\ a \triangleright m_2 = m_1,\ b \triangleright m_1 = -m_1,\ b \triangleright m_2 = -m_2$

(b) $a \triangleright m_1 = -m_2,\ a \triangleright m_2 = m_1,\ b \triangleright m_1 = -m_1,\ b \triangleright m_2 = -m_2$

(c) $a \triangleright m_1 = m_2,\ a \triangleright m_2 = m_1,\ b \triangleright m_1 = -m_1,\ b \triangleright m_2 = m_2$

(d) $a \triangleright m_1 = -m_2,\ a \triangleright m_2 = -m_1,\ b \triangleright m_1 = m_1,\ b \triangleright m_2 = m_2$

Exercise 5.6. Let $G = \langle g = (1\,2\,3), h = (1\,2) \mid g^3 = h^2 = \iota,\ hg = g^2 h \rangle \cong S_3$ and write $\zeta = e^{\frac{2\pi i}{3}} \in \mathbb{C}$.

For $\mathbb{K}$ a field and X a set, let $\mathrm{span}_{\mathbb{K}}\{X\}$ denote the $\mathbb{K}$-vector space spanned by elements of X.

Which of the following makes the given vector space M into a $\mathbb{C}[G]$-module?

(a) $M = \mathrm{span}_{\mathbb{C}}\{m\},\ g \triangleright m = m$ and $h \triangleright m = -m$.

(b) $M = \mathrm{span}_{\mathbb{C}}\{m_1, m_2\},\ g \triangleright m_1 = \frac{1}{2}m_1 + \frac{\sqrt{3}}{2}m_2,\ g \triangleright m_2 = -\frac{\sqrt{3}}{2}m_1 - \frac{1}{2}m_2,$
$h \triangleright m_1 = m_1$ and $h \triangleright m_2 = m_2$.

(c) $M = \mathrm{span}_{\mathbb{C}}\{m_1, m_2\},\ g \triangleright m_1 = \zeta m_1,\ g \triangleright m_2 = \zeta^2 m_2,\ h \triangleright m_1 = m_2$ and $h \triangleright m_2 = m_1$.

(d) $M = \mathrm{span}_{\mathbb{C}}\{m_1, m_2\},\ g \triangleright m_1 = \zeta m_1,\ g \triangleright m_2 = \zeta^2 m_2,\ h \triangleright m_1 = -m_1$ and $h \triangleright m_2 = -m_2$.

(e) $M = \mathrm{span}_{\mathbb{C}}\{m_1, m_2, m_3\},\ g \triangleright m_j = m_{j+1}$ for $j = 1, 2,\ g \triangleright m_3 = m_1$ and $h \triangleright m_j = m_j$ for $1 \leq j \leq 3$.

(f) $M = \mathrm{span}_{\mathbb{C}}\{m_1, m_2, m_3\},\ g \triangleright m_1 = m_3,\ g \triangleright m_j = m_{j-1}$ for $j = 2, 3$ and $h \triangleright m_j = v_{4-j}$ for $1 \leq j \leq 3$.

Exercise 5.7 (Problem 29). Let A be a $\mathbb{K}$-algebra. Show that the centre $Z(A)$ is a subalgebra of A.

Exercise 5.8. Find a basis for $Z(\mathbb{K}[G])$ and hence compare $\dim \mathbb{K}[G]$ and $\dim Z(\mathbb{K}[G])$ for

(a) $G = C_4$, the cyclic group of order 4;

(b) $G = D_8$, the dihedral group of order 8;

(c) $G = A_4$, the alternating group of degree 4.

Exercise 5.9. Let $G = \langle a \mid a^4 = e \rangle$ be a cyclic group of order 4 and M the $\mathbb{C}[G]$-module with basis $\mathcal{B} = \{m_1, m_2, m_3\}$ on which a acts by

$$a \triangleright m_1 = m_2,\ a \triangleright m_2 = -m_1,\ a \triangleright m_3 = m_1 + m_2 + m_3.$$

Find non-zero $\mathbb{C}[G]$-submodules M_1, M_2 and M_3 of M such that $M = M_1 \oplus M_2 \oplus M_3$.

Exercise 5.10. Let $G = \langle a \mid a^3 = e \rangle$ be the cyclic group of order 3 and let M be the $\mathbb{C}[G]$-module with basis $\mathcal{B} = \{m_1, m_2, m_3\}$ and action defined by

$$a \triangleright m_1 = m_2,\ a \triangleright m_2 = -m_1 - m_2,\ a \triangleright m_3 = m_2 + m_3.$$

Find all the 1-dimensional $\mathbb{C}[G]$-submodules of M.

Exercise 5.11. Let $G = D_{12} = \langle a, b \mid a^6 = b^2 = e,\ bab = a^{-1} \rangle$ be the dihedral group of order 12.

(a) Define

$$A = \frac{1}{2}\begin{pmatrix} 1 & 0 & -\sqrt{3} \\ 0 & 2 & 0 \\ \sqrt{3} & 0 & 1 \end{pmatrix}$$

and

$$B = \begin{pmatrix} 1 & 0 & 0 \\ 0 & 1 & 0 \\ 0 & 0 & -1 \end{pmatrix}$$

Let M be the 3-dimensional $\mathbb{C}[D_{12}]$-module with basis $\mathcal{B} = \{m_1, m_2, m_3\}$ for which the actions of a and b on $\mathcal{B}$ are given by A and B respectively (i.e. $a \triangleright m_1 = \frac{1}{2}m_1 + \frac{\sqrt{3}}{2}m_3$ etc.).

Find a 1-dimensional submodule M_1 of M. Is it simple?

(b) You are told that M/M_1 is simple. (If you wish, you can try to verify this now or later, when we have some more technology available.)

Hence give a composition series for M.

Exercise 5.12. Let $G = \langle a \mid a^3 = e \rangle$ be a cyclic group of order 3 and M, N two $\mathbb{C}[G]$-modules of dimension 2 with bases $\mathcal{B}_M = \{m_1, m_2\}$ and $\mathcal{B}_N =$

$\{n_1, n_2\}$ respectively. Suppose that the action of a on M and N is given by the matrices

$$[a]_{\mathcal{B}_M} = \begin{pmatrix} 9 & -7 \\ 13 & -10 \end{pmatrix}$$

and

$$[a]_{\mathcal{B}_N} = \begin{pmatrix} -5 & -3 \\ 7 & 4 \end{pmatrix}$$

Find $\lambda, \mu \in \mathbb{C}$ such that the map $\varphi \colon M \to N$ given by $\varphi(m_1) = -6n_1 + \lambda n_2$ and $\varphi(m_2) = -3n_1 + \mu n_2$ is a $\mathbb{C}[G]$-homomorphism.

[*With some thought, this can be done with computer assistance, but you should ensure that you can turn the output of any computations into a rigorous argument.*]

Exercise 5.13. Let G be an Abelian group and $\rho \colon G \to \mathrm{GL}_n(\mathbb{C})$ a representation of G. Write M for the corresponding $\mathbb{C}[G]$-module, and

$$E_\lambda(g) = \{m \in M \mid \rho(g)m = \lambda m\}$$

for the λ-eigenspace of $\rho(g)$ in M, for any $g \in G$ and $\lambda \in \mathbb{C}$. Prove the following statements:

(a) For any $g \in G$, every eigenspace of $\rho(g)$ is a $\mathbb{C}[G]$-submodule of M.

(b) ρ is faithful if and only if there is no non-identity element $g \in G$ such that $E_1(g) = M$.

Exercise 5.14 (Problem 30). Let $G = \langle a \mid a^4 = e \rangle = C_4$ be a cyclic group of order 4 and let M be the $\mathbb{C}[G]$-module of dimension 4 with basis $\mathcal{B} = \{m_1, m_2, m_3, m_4\}$ with

$$a \triangleright m_1 = m_2, \ a \triangleright m_2 = m_3, \ a \triangleright m_3 = m_4, \ a \triangleright m_4 = m_1.$$

(a) Show that $M_1 = \mathrm{span}_\mathbb{C}\{m_1 + m_2 + m_3 + m_4\}$ is a 1-dimensional $\mathbb{C}[G]$-submodule of M.

(b) Find three other 1-dimensional $\mathbb{C}[G]$-submodules M_2, M_3 and M_4 of M such that $M = M_1 \oplus M_2 \oplus M_3 \oplus M_4$.

[*Hint: consider vectors of the form $m_1 + \lambda m_2 + \lambda^2 m_3 + \lambda^3 m_4$ for suitable values of λ.*]

Exercise 5.15.

(a) Let N be a normal subgroup of a finite group G and let $\rho \colon G/N \to \mathrm{GL}_n(\mathbb{C})$ be a *faithful* representation of the quotient group G/N. Prove that the function

$$\widehat{\rho} \colon G \to \mathrm{GL}_n(\mathbb{C}), \ \widehat{\rho}(g) = \rho(gN)$$

(where $gN \in G/N$ is the coset containing g) is a representation of G with kernel N.

(b) Conversely, prove that any representation $\rho\colon G \to \mathrm{GL}_n(\mathbb{C})$ with kernel $K = \mathrm{Ker}\,\rho$ defines a faithful representation of the quotient group $\bar{\rho}\colon G/K \to \mathrm{GL}_n(\mathbb{C})$ with $\bar{\rho}(gK) = \rho(g)$ for any $g \in gK$ and any $gK \in G/K$.

Exercise 5.16. Let A be a finite-dimensional $\mathbb{K}$-algebra and let $_AA = (A, m)$ be the regular A-module (Example 4.1.12) with underlying vector space A and action $a \triangleright b = ab$ for all $a, b \in A$.

(a) Show that any A-module homomorphism $\varphi\colon {}_AA \to {}_AA$ is determined by $\varphi(1_A)$.

(b) Hence give a basis for $\mathrm{End}_A(_AA) = \mathrm{Hom}_{A\text{-Mod}}(_AA, {}_AA)$.

(c) (Hard!) Let $B = (B, m)$ be an algebra. The algebra B^{op} is defined to have the same underlying vector space B and multiplication map defined by $m^{\mathrm{op}}(b \otimes c) = m(c \otimes b)$; that is, in the opposite algebra B^{op}, the order of multiplication is reversed.

Prove that $A \cong \mathrm{End}_A(_AA)^{\mathrm{op}}$ as algebras.

Exercise 5.17. Let $G = \langle a \mid a^4 = e \rangle$ be a cyclic group of order 4 and let $A = \mathbb{C}[C_4]$.

(a) Let M be the 3-dimensional vector space with basis $\mathcal{B}_M = \{m_1, m_2, m_3\}$ and define
$$a \triangleright m_1 = -m_1 + (1 + \mathrm{i})m_2$$
$$a \triangleright m_2 = \mathrm{i}m_2$$
$$a \triangleright m_3 = (1 - \mathrm{i})m_1 + (-1 + 3\mathrm{i})m_2 - \mathrm{i}m_3$$
Show that $\triangleright$ extends to a map $\triangleright\colon A \otimes M \to M$ so that M becomes an A-module.

(b) Decompose M into a direct sum of simple submodules.

(c) Using Schur's lemma and related results, find a basis for $\mathrm{End}_A(M)$.

Exercise 5.18. Prove Lemma 5.2.14.

Exercise 5.19. [*This exercise shows that Maschke's theorem does not hold for infinite groups.*]

Let G be the additive group $(\mathbb{Z}, +)$ and let M be the $\mathbb{C}[G]$-module with basis $\mathcal{B} = \{m_1, m_2\}$ and action defined by
$$n \triangleright m_1 = m_1$$
$$n \triangleright m_2 = nm_1 + m_2$$
for each $n \in \mathbb{Z}$. Show that M has a simple submodule N of dimension 1 but N has no direct complement in M, i.e. there does not exist a (simple) submodule P of dimension 1 such that $M \cong N \oplus P$. That is, M is not semisimple.

Exercise 5.20. Let $G = \langle a \mid a^8 = e \rangle$ be a cyclic group of order 8 and let M be the $\mathbb{C}[G]$-module of dimension 2 with basis $\mathcal{B} = \{m_1, m_2\}$ and action defined by

$$a \triangleright m_1 = 7m_1 + 10m_2, \ a \triangleright m_2 = -5m_1 - 7m_2.$$

(a) Show that $M \cong N_1 \oplus N_2$ for $N_1 = \text{span}_{\mathbb{C}}\{n_1\}$ and $N_2 = \text{span}_{\mathbb{C}}\{n_2\}$ with $n_1 = (7 + \mathrm{i})m_1 + 10m_2$ and $n_2 = (7 - \mathrm{i})m_1 + 10m_2$.

(b) Give a composition series for M, justifying your answer.

(c) Using Schur's lemma and related results (which may include earlier exercises), find a basis for $\text{End}_{\mathbb{C}[G]}(M)$.

Exercise 5.21. Let G be a finite group and M a $\mathbb{C}[G]$-module. Define

$$M_f = \{m \in M \mid g \triangleright m = m \ \forall g \in G\}$$

(a) Prove that M_f is a $\mathbb{C}[G]$-submodule of M.

(b) Prove that for the regular module $M = {}_{\mathbb{C}[G]}\mathbb{C}[G]$ we always have $\dim M_f \geq 1$. That is, show there is a non-zero element m of ${}_{\mathbb{C}[G]}\mathbb{C}[G]$ such that $g \triangleright m = m$ for all $g \in G$.

(c) Let N be a $\mathbb{C}[G]$-module. Show that $M \cong N$ implies $M_f \cong N_f$.

Exercise 5.22 (Problem 31). Let n be an integer $n \geq 2$, let $G = \langle a \mid a^n = e \rangle$ be a cyclic group of order n and let M be the permutation $\mathbb{C}[G]$-module of dimension n with natural basis $\mathcal{B} = \{m_1, \ldots, m_n\}$ defined by

$$a \triangleright m_i = m_{i+1}$$

where we interpret the indices modulo n. Find all the 1-dimensional $\mathbb{C}[G]$-submodules of M.

Exercise 5.23 (cf. Problem 32). Let G be a group of order 12 and $\{S_1, \ldots, S_r\}$ be a complete set of non-isomorphic simple $\mathbb{C}[G]$-modules. List all the possible integers r and hence all the sequences of dimensions

$$(\dim S_1, \dim S_2, \ldots, \dim S_r)$$

Do the same for G a group of order 18.

Exercise 5.24. Prove Lemma 5.3.1.

Exercise 5.25. Using Example 5.3.11 as a template, write down the simple representations of the linearly oriented A_4 quiver

$$1 \longrightarrow 2 \longrightarrow 3 \longrightarrow 4$$

Exercise 5.26 (Problem 33). Let $\mathbb{K}$ be a field and A a $\mathbb{K}$-algebra. If $M = (M, \triangleright)$ is a left A-module, then M^* is naturally a right A-module. If $(M, \triangleleft)$ is a right A-module, then M^* is naturally a left A-module.

[*Consider* $(f \triangleleft a)(m) \overset{\text{def}}{=} f(a \triangleright m)$.]

Exercise 5.27. Using Proposition 5.3.19, write down the simple, (indecomposable) projective and injective modules, and identify any isomorphisms between them, for the quivers (i) (linearly oriented) A_3 and (ii) (linearly oriented) A_4.

For a challenge, choose your own acyclic quiver and do the same.

Chapter 6

⑤ Advanced topics

6.1 ⑤ The Lasker–Noether theorem

In this section, we generalize the cyclic decomposition theorem of Section 5.1 to the case of Noetherian rings. The condition of a ring or module being Noetherian (or its dual, Artinian) imposes a certain level of finiteness: it allows for growth but not in an uncontrollable way. As such, many people treat Noetherianity as a base condition, so that one has at least a few tools to use and exotic examples are put to one side from the start.

Definition 6.1.1. Let R be a ring and M a left R-module. We say that M is (left) *Noetherian* if M has the *ascending chain condition* (ACC) on its submodules. That is, for every sequence

$$M_0 \leq M_1 \leq M_2 \leq \cdots$$

of submodules of M, there exists $n \in \mathbb{N}$ such that $M_n = M_{n+i}$ for all $i \in \mathbb{N}$.

Provided we allow the axiom of choice, this is equivalent to each of the following two statements:

- any non-empty set S of submodules of M has at least one maximal element with respect to inclusion;

- every submodule of M is finitely generated.

The second condition makes it clear that a Noetherian module is itself finitely generated. The point here is that, in general, non-Noetherian modules can be finitely generated but have submodules that are not finitely generated.

Indeed, given a module M and submodule K, one can show (exercise) that M is Noetherian if and only if K and M/K are Noetherian, so that Noetherianity plays nicely with extensions. In particular, this is true for split extensions: if M and N are Noetherian, so is $M \oplus N$.

Specializing to the regular module $_R R$, we have:

Definition 6.1.2. Let R be a ring. We say that R is *left Noetherian* if $_R R$ is Noetherian.

©2025 Jan E. Grabowski, CC BY-NC 4.0 https://doi.org/10.11647/OBP.0492.06

If R is commutative, being left Noetherian is equivalent to being right Noetherian (i.e. R_R is right Noetherian), but again in general it is possible to be one and not the other. Since the left submodules of $_RR$ are exactly the left ideals, we have that being a left Noetherian ring is equivalent to the above two statements with "submodule" replaced by "left ideal". Note that the stronger condition of being an ideal is important; Noetherian rings can certainly have non-Noetherian subrings.

Some examples are as follows; in some cases, the proofs are quite non-trivial.

Examples 6.1.3.

(a) Fields are Noetherian; this is straightforward as there are only two ideals to consider.

(b) A principal ideal domain (PID) is Noetherian: every left ideal is generated by a single element, so certainly by finitely many! (In this way, one can see Noetherianity as a generalization of being a PID, which motivates this section.)

(c) The ring of polynomials in finitely many variables over a Noetherian ring is again Noetherian; this is the celebrated Hilbert basis theorem.

Note that polynomial rings in infinitely many variables are not Noetherian (see the above opening remarks on finiteness) and rings of continuous functions, e.g. $\mathcal{C}(\mathbb{R}, \mathbb{R})$, are rarely Noetherian.

To tie back to earlier discussions on types of modules, we note too the following.

Theorem 6.1.4 (Bass). *A ring R is left Noetherian if and only if every direct sum of injective R-modules is injective.*

Compare this with the remarks after Proposition 4.5.4, where we saw that direct sums of projectives are projective (always).

Rather than having to check that each module over a ring R is Noetherian one by one, we have the following, saying that if our ring R is Noetherian, so is every finitely generated module over it. This generalizes the claim of Proposition 5.1.8.

Proposition 6.1.5. *Every finitely generated R-module over a Noetherian ring is Noetherian.*

Proof. From the above, one has that finitely generated free modules over Noetherian rings are Noetherian, being a direct sum of copies of $_RR$.

Now if M is a finitely generated module, we have $\pi \colon {_RR^{\oplus n}} \twoheadrightarrow M$ so that $M \cong {_RR^{\oplus n}}/\operatorname{Ker}\pi$. So M is a quotient of a Noetherian module and hence Noetherian. $\qquad\square$

We conclude by stating the Lasker–Noether theorem, first in its most general form for modules, then in its more common form for rings.

We need the notion of a *primary* submodule of a module: we say $N \leq M$ is primary if for any $r \in R$ a zero-divisor on M/N (i.e. for which there exists $\bar{n} \in M/N$ such that $r \triangleright \bar{n} = 0$), r acts nilpotently on M/N (i.e. there exists $k \in \mathbb{N}$ such that $x^k \triangleright M/N = 0$).

Theorem 6.1.6 (Lasker–Noether theorem for modules). *Let R be a Noetherian ring. Then every submodule of a finitely generated R-module is a finite intersection of primary submodules.*

Theorem 6.1.7 (Lasker–Noether theorem for rings). *Let R be a Noetherian ring. Then every left ideal of R is a finite intersection of primary ideals.*

Since PIDs are Noetherian, with a little work to line things up, this recovers the cyclic decomposition theorem, Theorem 5.1.5. For $\mathbb{Z}$ in particular, the module version recovers the classification of finitely generated Abelian groups (Corollary 5.1.10) and the ring version is in fact the Fundamental Theorem of Arithmetic, i.e. the factorization of integers into primes.

In geometry, this result is important too: by application to polynomial rings in finitely many variables over a field, it says that algebraic sets may be decomposed into a finite union of irreducible varieties.

The interested reader can find details and much more (including the non-Noetherian case) in [AM].

6.2   🍋 Auslander–Reiten theory

Our goal in this section is to give a brief introduction to an important tool in the study of representations of finite-dimensional algebras. The underpinning theory is due to Auslander and Reiten ([AR]), so is known as Auslander–Reiten theory, or often "AR theory" for short.

The key insight of AR theory is that, in many cases, to completely describe the finite-dimensional representation theory, A-mod, of a finite-dimensional $\mathbb{K}$-algebra A, it suffices to know two classes of data:

- the indecomposable modules; and

- the irreducible morphisms between indecomposables.

The irreducible morphisms may be combined into special short exact sequences (called *almost split* sequences), giving us an explicit description of (most of) A-mod.

We will not give full details, just the key definitions and some examples; starting points for a more in-depth introduction are [ASS], [ARS] and [Sch], according to one's taste.

In contrast to the rest of this book, we will now also restrict to considering finite-dimensional modules over a $\mathbb{K}$-algebra A, which will itself be assumed to be finite-dimensional. We will denote by A-mod the full[1] subcategory of A-Mod whose objects are the finite-dimensional A-modules.

In Section 4.4, we defined indecomposability for modules. The other key definition for AR theory is that of an irreducible morphism.

Definition 6.2.1. Let $L, N \in A$-mod and let $f: L \to N$ be a morphism of A-modules. We say f is *irreducible* if

(a) f is neither a section nor a retraction; and

(b) if f factors through $M \in A$-mod, i.e. there exists $g: L \to M$ and $h: M \to N$ such that $f = h \circ g$, then either g is a section or h is a retraction.

Unfortunately, the irreducible morphisms do not form a subspace of $\mathrm{Hom}_{A\text{-mod}}(L, N)$, since the zero morphism is not irreducible. However, the set $\mathrm{NIrr}(L, N)$ of *non*-irreducible morphisms between indecomposables L and N is a subspace, and we heavily abuse notation by defining "$\dim \mathrm{Irr}(L, N)$" to be the result of calculating $\dim \mathrm{Hom}_{A\text{-mod}}(L, N) - \dim \mathrm{NIrr}(L, N)$; we apologise for this (see Remark 6.2.3 below).

Definition 6.2.2. A short exact sequence in A-mod

$$0 \longrightarrow L \overset{f}{\longhookrightarrow} M \overset{g}{\longtwoheadrightarrow} N \longrightarrow 0 \tag{6.1}$$

is said to be *almost split* if L and N are indecomposable and f and g are irreducible morphisms.

If (6.1) is almost split, we say that L is the *Auslander–Reiten translation* of N and write $L = \tau N$.

Remark 6.2.3. We are making these definitions for expediency, to allow a compact and self-contained exposition. However, this is not the "right" way to do things. There is a much better definition of $\mathrm{Irr}(L, N)$ in terms of powers of radicals; see [ASS]. Also, as explained in [Sch], one should define left and right almost split morphisms and thence almost split sequences, at which point the first part of Definition 6.2.2 becomes a lemma. There is also a much more general functorial construction of Auslander–Reiten translation, so that the second part also becomes a lemma.

This notwithstanding, our short-term goal is the following.

Definition 6.2.4. Let A be a finite-dimensional $\mathbb{K}$-algebra. The *Auslander–Reiten (AR) quiver* of A is the quiver with vertices the isomorphism classes[2] of indecomposable (finite-dimensional) A-modules and $\dim \mathrm{Irr}(L, N)$ arrows from L to N.

---

[1]A full subcategory $\mathcal{D}$ of a category $\mathcal{C}$ is given by specifying the objects of $\mathcal{D}$, as a subcollection of $\mathrm{Obj}(\mathcal{C})$, and then setting $\mathrm{Hom}_{\mathcal{D}}(D, E) = \mathrm{Hom}_{\mathcal{C}}(D, E)$ for all $D, E \in \mathrm{Obj}(\mathcal{D})$, i.e. taking all morphisms from $\mathcal{C}$ as the morphisms between objects of $\mathcal{D}$.

[2]Isomorphism of modules is an equivalence relation and isomorphism classes are the equivalence classes with respect to this. We often say "isoclasses" for short; every second saved by not saying "morphism" is another second to understand more maths, after all.

We will give some examples of AR quivers shortly in Examples 6.2.6, but first record some facts from AR theory, to illustrate the type of information we get.

Proposition 6.2.5. *Let A be a finite-dimensional $\mathbb{K}$-algebra and M, N indecomposable finite-dimensional A-modules.*

(i) *N is projective if and only if $\tau N = 0$;*

(ii) *if N is not projective then τN is indecomposable;*

(iii) *if M, N are not projective then $M \cong N$ if and only if $\tau M \cong \tau N$; and*

(iv) *(the Auslander–Reiten formulæ)*

$$\mathrm{Ext}^1(M,-) \cong D\overline{\mathrm{Hom}}(-,\tau M)$$
$$\mathrm{Ext}^1(-,N) \cong D\overline{\mathrm{Hom}}(N,\tau-)$$

where $D = \mathrm{Hom}_{\mathbb{K}\text{-Mod}}(-,\mathbb{K})$ is $\mathbb{K}$-duality and

$$\overline{\mathrm{Hom}}(X,Y) = \mathrm{Hom}_{A\text{-Mod}}(X,Y)/I(X,Y)$$

for $I(X,Y)$ the subspace of all morphisms $f \colon X \to Y$ that factor through an injective A-module.

Examples 6.2.6.

(a) Recall Example 5.3.28(a): for the A_2 quiver $1 \overset{\alpha}{\longrightarrow} 2$ we found three indecomposable modules (up to isomorphism)—S_1, I_2 and S_2—and saw that these sit in a short exact sequence

$$0 \longrightarrow S_2 \hookrightarrow I_2 \twoheadrightarrow S_1 \longrightarrow 0$$

In fact, this sequence is almost split, so $S_2 = \tau S_1$. We can draw the AR (Auslander–Reiten) quiver for this quiver as

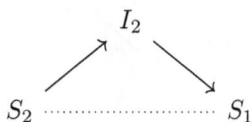

where the dotted line indicates Auslander–Reiten translation (that is, $\tau M \cdots\cdots M$). This is an example of a *mesh*, which is a particular subquiver of the AR quiver that records the existence of an almost split sequence; we will see more complicated meshes below.

(b) Consider again the (acyclic) linearly oriented quiver of type A_3, $1 \xrightarrow{\alpha} 2 \xrightarrow{\beta} 3$. Using the information about simple, projective and injective indecomposables from Proposition 5.3.19, we can derive that the AR quiver is as follows:

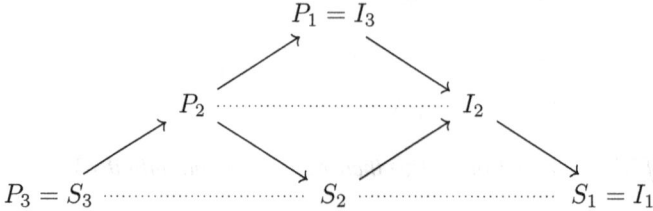

Recalling that $\tau M = 0$ if and only if M is projective, we expect to see these on the left of the AR quiver and indeed, there they are. Dually, the injectives appear on the right edge.

There are two meshes (with bases the simples) that are as in the A_2 cases, but also one other mesh, corresponding to the following almost split sequence:

$$0 \longrightarrow P_2 \hookrightarrow P_1 \oplus S_2 \twoheadrightarrow I_2 \longrightarrow 0 \qquad (6.2)$$

Using the AR quiver (and as explained in detail in [Sch, 3.1.4]) we can determine dimensions of Hom and Ext^1 spaces. For example, the function $\dim \mathrm{Hom}(P_2, -)$ has the following values

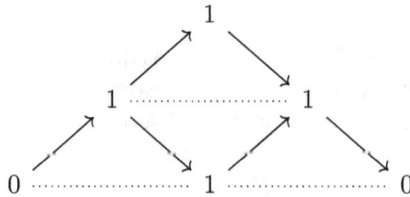

and $\dim \mathrm{Ext}^1(I_2, -) = \dim D\mathrm{Hom}(-, \tau I_2) = \dim D\mathrm{Hom}(-, P_2)$ is represented by

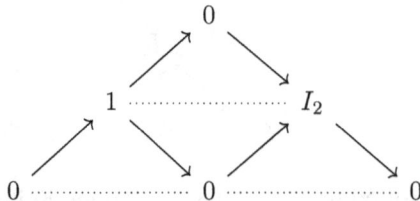

indicating that there is a single extension (up to isomorphism), as we should expect, given the existence of the sequence (6.2) with end terms P_2 and I_2.

(c) Changing the orientation does affect the representation theory and the AR quiver (although herein is a gateway to an interesting and important area of current research). The AR quiver for the quiver $1 \xrightarrow{\alpha} 2 \xleftarrow{\beta} 3$ whose underlying graph is still the Dynkin diagram of type A_3 is as follows:

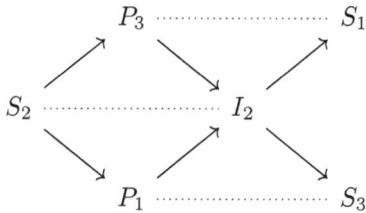

Note that several numerical aspects remain unchanged (e.g. the number of simples, projectives, injectives and—in fact—indecomposables) but others change.

(d) We finish with an example not of finite representation type, but rather of *tame* type. There is a technical definition of tame type[3] but it suffices here to say that tame, while not being of finite type, has sufficiently controlled infinite behaviour to still be able to say quite a lot about its representation theory. (The generic case is *wild* representation type—which, remarkably, turns out to mean[4] that one has all possible representation theory at the same time.)

We give the canonical tame example, the Kronecker quiver $1 \rightrightarrows 2$ over $\mathbb{K}$ algebraically closed.[5] Its AR quiver has the form

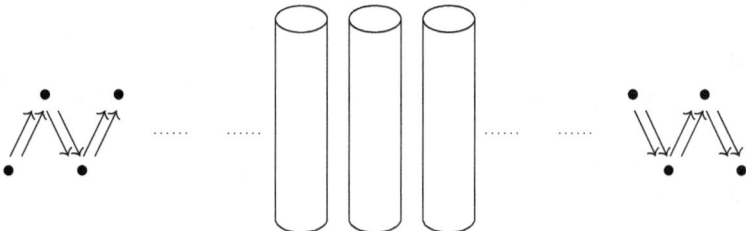

where the middle components here are known as *tubes*, for (hopefully) apparent reasons.

---

[3]Go find it!
[4]Paraphrasing somewhat...
[5]This now matters.

Tubes are made by gluing the two vertical edges of a quiver of the form

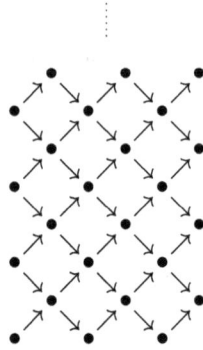

6.3 ⚐ Modules for Hopf algebras

In this section, we give a brief introduction to Hopf algebras. These tie together algebras—and in particular, group algebras—and monoidal categories, allowing us to explore how some of the theory of Section 5.2 can be generalized.

Recall from Section 4.1.3 our description of an associative unital $\mathbb{K}$-algebra in terms of maps on certain tensor products; for this section, we will only work over fields and will suppress the extra subscript $\mathbb{K}$ on the tensor product symbol. That is, we have that (A, m, u) is a $\mathbb{K}$-algebra if A is a $\mathbb{K}$-vector space (i.e. a $\mathbb{K}$-module) and $m\colon A \otimes A \to A$ and $u\colon \mathbb{K} \to A$ are $\mathbb{K}$-linear maps and these are such that the following diagrams commute:

(a) (associativity)

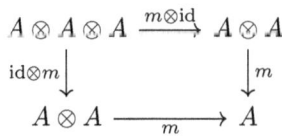

$$
\begin{array}{ccc}
A \otimes A \otimes A & \xrightarrow{\ m\otimes\mathrm{id}\ } & A \otimes A \\
{\scriptstyle \mathrm{id}\otimes m}\big\downarrow & & \big\downarrow{\scriptstyle m} \\
A \otimes A & \xrightarrow[\ m\ ]{} & A
\end{array}
$$

(b) (unitarity)

$$
\begin{array}{ccccc}
 & & A \otimes A & & \\
 & {\scriptstyle u\otimes\mathrm{id}}\nearrow & \big\uparrow{\scriptstyle m} & \nwarrow{\scriptstyle \mathrm{id}\otimes u} & \\
\mathbb{K} \otimes A & & & & A \otimes \mathbb{K} \\
 & {\scriptstyle \cong}\searrow & \big\downarrow & \swarrow{\scriptstyle \cong} & \\
 & & A & &
\end{array}
$$

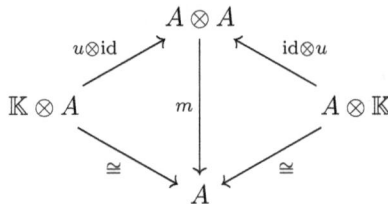

Here, the maps marked "$\cong$" are the canonical maps sending $\lambda \otimes a$ and $a \otimes \lambda$ to λa.

If we consider the $\mathbb{K}$-dual vector space $A^* \stackrel{\text{def}}{=} \text{Hom}_{\mathbf{Vect}_{\mathbb{K}}}(A, \mathbb{K})$, this is again a $\mathbb{K}$-vector space under pointwise addition and scalar multiplication of functionals. However it is not, in general, an algebra: when we dualize, the maps m and u do not carry over to induce an algebra structure. Rather, they induce the structure of a *coalgebra*.

Definition 6.3.1. Let C be a $\mathbb{K}$-vector space. We say that (C, Δ, ε) is a coalgebra if $\Delta: C \to C \otimes C$ and $\varepsilon: C \to \mathbb{K}$ are $\mathbb{K}$-linear maps such that the following diagrams commute:

(a) (coassociativity)

$$
\begin{array}{ccc}
C \otimes C \otimes C & \xleftarrow{\;\Delta \otimes \text{id}\;} & C \otimes C \\
\uparrow{\scriptstyle \text{id} \otimes \Delta} & & \uparrow{\scriptstyle \Delta} \\
C \otimes C & \xleftarrow{\;\;\Delta\;\;} & C
\end{array}
$$

(b) (counitarity)

$$
\begin{array}{ccccc}
& & C \otimes C & & \\
{\scriptstyle \varepsilon \otimes \text{id}} \swarrow & & \uparrow{\scriptstyle \Delta} & & \searrow {\scriptstyle \text{id} \otimes \varepsilon} \\
\mathbb{K} \otimes C & & \Delta & & C \otimes \mathbb{K} \\
& {\scriptstyle \cong} \nwarrow & \uparrow & \nearrow {\scriptstyle \cong} & \\
& & C & &
\end{array}
$$

The map Δ is called the *comultiplication* of the coalgebra C and ε the *counit*.

These conditions are not as familiar as associativity, say, and rather than dwell overly on them in full generality, we will see what happens in examples shortly.

We have a notion of cocommutativity, dual to that of commutativity. Recall that we have $\tau: C \otimes C \to C \otimes C$, $\tau(c_1 \otimes c_2) = c_2 \otimes c_1$ the natural map that 'flips' the tensor product. We say that a coalgebra (C, Δ, ε) is *cocommutative* if $\tau \circ \Delta = \Delta$.

First, we note that we have a natural notion of maps between coalgebras.

Definition 6.3.2. Let $(C, \Delta_C, \varepsilon_C)$ and $(D, \Delta_D, \varepsilon_D)$ be $\mathbb{K}$-coalgebras. A *homomorphism* of $\mathbb{K}$-coalgebras is a $\mathbb{K}$-linear map $f: C \to D$ such that the following diagrams commute:

(a) (comultiplication preserved)

$$
\begin{array}{ccc}
C \otimes C & \xrightarrow{\;f \otimes f\;} & D \otimes D \\
\uparrow{\scriptstyle \Delta_C} & & \uparrow{\scriptstyle \Delta_D} \\
C & \xrightarrow{\;\;f\;\;} & D
\end{array}
$$

(b) (counit preserved)

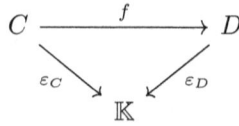

$$
\begin{array}{ccc}
C & \xrightarrow{\ f\ } & D \\
& \varepsilon_C \searrow \quad \swarrow \varepsilon_D & \\
& \mathbb{K} &
\end{array}
$$

Now, in certain special but important cases, we can have both algebra and coalgebra structures at once, in a compatible way, giving us some self-duality. It turns out that it is helpful to ask for one additional piece of structure relating the algebra and coalgebra to each other and this gives rise to the definition of a Hopf algebra.

Definition 6.3.3. Let (A, m, u) be a $\mathbb{K}$-algebra and let $\Delta\colon A \to A \otimes A$ and $\varepsilon\colon A \to \mathbb{K}$ be $\mathbb{K}$-algebra homomorphisms satisfying the coassociativity and counitarity conditions. Then we say that $(A, m, u, \Delta, \varepsilon)$ is a $\mathbb{K}$-bialgebra.

The compatibility of the algebra and coalgebra structures is encoded in the condition that Δ and ε are algebra homomorphisms, not just $\mathbb{K}$-linear maps. One can show that this is equivalent to requiring that m and u are coalgebra homomorphisms, so that there is not in fact the asymmetry that there first appears to be.

A homomorphism of bialgebras is a linear map that is both an algebra and coalgebra homomorphism.

If B is finite-dimensional then B^* is again a bialgebra, in the natural way, with the multiplication and comultiplication (and unit and counit) exchanged. (In the infinite-dimensional case, this is not true, and a notion of finite dual is needed.)

The extra structure we need for a Hopf algebra is a certain anti-automorphism, known as an *antipode*.

Definition 6.3.4. Let $H = (H, m, u, \Delta, \varepsilon)$ be a $\mathbb{K}$-bialgebra. Let $S\colon H \to H$ be a $\mathbb{K}$-linear map such that the following diagram commutes:

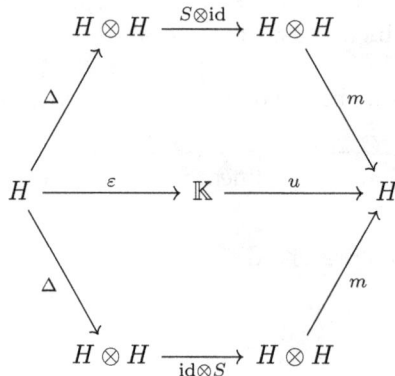

$$
\begin{array}{ccccc}
H \otimes H & \xrightarrow{\ S \otimes \mathrm{id}\ } & H \otimes H & & \\
\Delta \nearrow & & & \searrow m & \\
H & \xrightarrow{\ \varepsilon\ } & \mathbb{K} & \xrightarrow{\ u\ } & H \\
\Delta \searrow & & & \nearrow m & \\
& H \otimes H & \xrightarrow[\ \mathrm{id} \otimes S\ ]{} & H \otimes H &
\end{array}
$$

Then we say that $(H, m, u, \Delta, \varepsilon, S)$ is a *Hopf algebra* over $\mathbb{K}$.

We now give some examples, the first of which is the promised familiar one.

Examples 6.3.5.

(a) The group algebra $\mathbb{K}[G]$ is a Hopf algebra with $\Delta(g) = g \otimes g$, $\varepsilon(g) = 1_{\mathbb{K}}$ and $S(g) = g^{-1}$ for all $g \in G$.

   This form of comultiplication has a special name: we say that an element $h \in H$ is *group-like* if $\Delta(h) = h \otimes h$. The set of group-like elements of H forms a group under m, with the inverse of h being $S(h)$. In the group algebra $\mathbb{K}[G]$, taking the group of group-like elements recovers the group G.

   The group Hopf algebra is commutative if and only if G is Abelian but is always cocommutative: $(\tau \circ \Delta)(g) = \tau(g \otimes g) = g \otimes g = \Delta(g)$.

(b) If G is a finite group, the algebra $\mathbb{K}(G) = \mathrm{Hom}_{\mathbf{Set}}(G, \mathbb{K})$ is a Hopf algebra with $\Delta(f)(g, h) = f(gh)$, $\varepsilon(f) = f(e)$ and $S(f)(g) = f(g^{-1})$. (The definition of the coproduct here implicitly uses the isomorphism $\mathbb{K}(G \times G) \cong \mathbb{K}(G) \otimes \mathbb{K}(G)$.)

   This Hopf algebra is always commutative and is cocommutative if and only if G is Abelian.

(c) Let V be a vector space over $\mathbb{K}$. The *tensor algebra* $T(V) \stackrel{\mathrm{def}}{=} \bigoplus_{r \in \mathbb{N}} V^{\otimes r}$ is a Hopf algebra with

$$m((v_1 \otimes \cdots \otimes v_r) \otimes (w_1 \otimes \cdots \otimes w_s)) = v_1 \otimes \cdots \otimes v_r \otimes w_1 \otimes \cdots \otimes w_s,$$

   $u(1_{\mathbb{K}}) = 1_{\mathbb{K}} \in V^{\otimes 0} \equiv \mathbb{K}$ and $\Delta(v) = v \otimes 1 + 1 \otimes v$, $\varepsilon(v) = 0$ and $S(v) = -v$ all defined for $v \in V$; one can check that these specifications can be extended to $T(V)$ to give maps with the required properties.

   Elements $h \in H$ for which $\Delta(h) = h \otimes 1 + 1 \otimes h$ are called *primitive*; the set of primitive elements of a Hopf algebra becomes a *Lie algebra* with respect to the Lie bracket $[h, k] = hk - kh$.

   The *symmetric algebra* and *exterior algebra* over a vector space V are Hopf quotients of $T(V)$, so Hopf algebras in their own right.

There are many more examples arising in *quantum algebra and noncommutative geometry*. The interested reader can find more in [Maj] and [BG].

The representation theory of Hopf algebras has some nice features, beyond what one can expect without the extra structure. Let H be a Hopf algebra over $\mathbb{K}$. Then H-Mod is a monoidal category (2.7) via Δ. For if $(M, \rhd_M), (N, \rhd_N) \in H$-Mod, we can define $\rhd \colon H \otimes M \otimes N \to M \otimes N$ by

$$\rhd = (\rhd_M \otimes \rhd_N) \circ (\mathrm{id} \otimes \tau \otimes \mathrm{id}) \circ (\Delta \otimes \mathrm{id} \otimes \mathrm{id})$$

and so make $M \otimes N$ an H-module. More concretely, if $\Delta(h) = \sum_i h_{1i} \otimes h_{2i} \in H \otimes H$, we have $h \rhd (m \otimes n) = \sum_i (h_{1i} \rhd_M m) \otimes (h_{2i} \rhd_N n)$.

Hopf algebras always have a trivial module, the vector space $\mathbb{K}$ with action $\triangleright\colon H \otimes \mathbb{K} \to \mathbb{K}$, $h \triangleright \lambda = \varepsilon(h)\lambda$.

Finally, we always have duals: if $(M, \triangleright_M) \in H$-Mod, then its $\mathbb{K}$-dual $M^* = \mathrm{Hom}_{\mathbf{Vect}_\mathbb{K}}(M, \mathbb{K})$ is an H-module via $(h \triangleright f)(m) = f(S(h) \triangleright m)$ for all $m \in M$.

The name for monoidal categories having these extra properties is *rigid*: the category of modules over a Hopf algebra H is a rigid monoidal category equipped with a fibre functor, the forgetful functor $\mathcal{F}\colon H$-Mod $\to \mathbb{K}$-Mod $\equiv$ **Vect**$_\mathbb{K}$, and the main theorem of *Tannaka duality* is that one can reconstruct H from the endomorphisms of the functor $\mathcal{F}$. This is closely related to the Yoneda lemma (2.5) in contexts with only algebraic structures around but is much deeper for examples arising from harmonic analysis. It is a far-reaching generalization of Problem 26 and we give a related example in the next section.

6.4 ⑤ Representations of categories

Earlier, we defined a representation of a quiver via its path category. We can generalize this and other examples, such as representations of groups, to *representations of categories*.

Definition 6.4.1. Let $\mathcal{C}$ be a category and $\mathbb{K}$ a field. A $\mathbb{K}$-*linear representation* of $\mathcal{C}$ is a functor $F\colon \mathcal{C} \to \mathbb{K}$-Mod.

Examples 6.4.2.

(a) Let G be a group and $\mathcal{G}$ the 1-object groupoid[6] on $*$ with $\mathrm{Hom}_\mathcal{G}(*, *) = G$. Then a representation of $\mathcal{G}$ is a choice of vector space $V = F* \in \mathbb{K}$-Mod and linear maps $Fg \in \mathrm{Hom}_{\mathbb{K}\text{-Mod}}(V, V)$ for each $g \in G$ such that $Fh \circ Fg = F(hg)$. Since G is a group, Fg is invertible (with inverse Fg^{-1}) so $Fg \in \mathrm{End}_{\mathbb{K}\text{-Mod}}(V) = \mathrm{GL}_\mathbb{K}(V)$ and $g \mapsto Fg$ is a group homomorphism. That is, we have recovered the definition of a linear group representation.

(b) As noted above, a representation of a quiver is by definition a representation of the path category of the quiver.

These examples give a good sense of what happens in general: to each object c of $\mathcal{C}$ the representation F associates a $\mathbb{K}$-vector space Fc, and to each morphism $f\colon c \to d$, F associates a linear map $Ff\colon Fc \to Fd$, in a way that is compatible with composition of morphisms.

As one might expect by now, we can define morphisms between representations of categories and hence a category of representations.

---

[6]A *groupoid* is a category in which every morphism is invertible. One-object groupoids are determined by the automorphism group of their object so are in 1-1 correspondence with groups. Groupoids with more than one object are of interest in a number of settings, especially algebraic topology in the construction of the *fundamental groupoid* of a space.

Definition 6.4.3. Let $\mathcal{C}$ be a category, $\mathbb{K}$ a field and let $F, G \colon \mathcal{C} \to \mathbb{K}$-Mod be $\mathbb{K}$-linear representations of $\mathcal{C}$. A *morphism* of representations is a natural transformation of functors, $\alpha \colon F \Rightarrow G$.

This definition has the advantage of being both entirely natural in the category theory world and of encoding a suite of somewhat complicated conditions in one go. A helpful exercise we recommend is to compare this definition in the case of $\mathcal{C} = \mathcal{G}$ a 1-object groupoid with Definition 3.1.17.

We immediately know how to define an isomorphism of representations, namely as a natural isomorphism of functors. Indeed, all this is taking place within the *functor category* $[\mathcal{C}, \mathbb{K}\text{-Mod}]$.

Definition 6.4.4. Let $\mathcal{C}$ be a category and $\mathbb{K}$ a field. The category of representations of $\mathcal{C}$ is the functor category $\mathrm{Rep}(\mathcal{C}) \overset{\text{def}}{=} [\mathcal{C}, \mathbb{K}\text{-Mod}]$ with objects functors from $\mathcal{C}$ to $\mathbb{K}$-Mod and morphisms natural transformations of functors.

Since $\mathbb{K}$-Mod has a lot of structure—we have seen that it is both Abelian (4.2.1) and also rigid monoidal (2.7,6.3)—the category of representations inherits a lot of structure too, by *pulling back* from $\mathbb{K}$-Mod.

Proposition 6.4.5. *Let $\mathcal{C}$ be a category and $\mathbb{K}$ a field. Then $\mathrm{Rep}(\mathcal{C})$ is Abelian.*

Proof (sketch): Given two morphisms $\alpha, \beta \colon F \Rightarrow G$, we can define their sum by $(\alpha+\beta)_c \overset{\text{def}}{=} \alpha_c + \beta_c$; the latter is well-defined since $\alpha_c, \beta_c \in \mathrm{Hom}_{\mathbf{Vect}_{\mathbb{K}}}(Fc, Gc)$ and $\mathbb{K}$-Mod is Abelian. Indeed pulling back from $\mathbb{K}$-Mod, we have that $\mathrm{Rep}(\mathcal{C})$ is **Ab**-enriched.

The zero object in $\mathrm{Rep}(\mathcal{C})$ is given by the functor $\mathbf{0} \colon \mathcal{C} \to \mathbb{K}\text{-Mod}$, $\mathbf{0}c = 0$ for all $c \in \mathcal{C}$ and $\mathbf{0}f = 0$ for all $f \colon c \to d$.

Similarly, we can define both finite Cartesian products and finite direct sums object by object, e.g. $F \oplus G \colon \mathcal{C} \to \mathbb{K}\text{-Mod}$ is given by $(F \oplus G)c = Fc \oplus Gc$ and $(F \oplus G)f = Ff \oplus Gf$.

Taking kernels and cokernels locally, we have these for $\mathrm{Rep}(\mathcal{C})$: indeed, for $\alpha \colon F \Rightarrow G$ we have $\mathrm{Ker}\,\alpha \colon F \Rightarrow G$ defined by $(\mathrm{Ker}\,\alpha)_c \overset{\text{def}}{=} \mathrm{Ker}\,\alpha_c$ and similarly for cokernels. Similarly, if $\alpha \colon F \Rightarrow G$ is a monomorphism, we can form $Gc/\mathrm{Im}\,\alpha_c \cong Fc$ in $\mathbb{K}$-Mod for all c to see that α is a kernel, and dually for cokernels. $\qquad\square$

Proposition 6.4.6. *Let $\mathcal{C}$ be a category and $\mathbb{K}$ a field. Then $\mathrm{Rep}(\mathcal{C})$ is a monoidal category with respect to $F \otimes G \colon \mathcal{C} \to \mathbb{K}\text{-Mod}$ defined by $(F \otimes G)c = Fc \otimes Gc$ and $(F \otimes G)f = Ff \otimes Gf$.*

In some instances, we have natural representations to consider. For example, if A is a $\mathbb{K}$-algebra, the forgetful functor $\mathcal{R} \colon A\text{-Mod} \to \mathbb{K}\text{-Mod}$ is a representation of A-Mod.

Proposition 6.4.7. *Let A be a $\mathbb{K}$-algebra. The forgetful functor $\mathcal{R} \colon A\text{-Mod} \to \mathbb{K}\text{-Mod}$ has $\mathrm{End}_{\mathrm{Rep}(A\text{-Mod})}(\mathcal{R})$ isomorphic to A as $\mathbb{K}$-modules.*

Proof. The functor $\mathcal{R}$ is representable: it forms an adjoint pair with a functor Free: $\mathbb{K}$-Mod $\rightarrow$ A-Mod, as a version of the forgetful-free adjunction but for $\mathbb{K}$-Mod-enriched (i.e. $\mathbb{K}$-linear) categories and functors. In particular, $\mathcal{R} \cong \text{Hom}_{A\text{-Mod}}(_AA, -)$ is represented by the regular module; in less fancy terms, this is the claim that for any A-module M, $\text{Hom}_{A\text{-Mod}}(_AA, M) \cong M$ as vector spaces, via $f \mapsto f(1_A)$ (either check for yourself or see [DK, §1.7]). Now

$$E = \text{End}_{\text{Rep}(A\text{-Mod})}(\mathcal{R}) \cong \text{Hom}_{\text{Rep}(A\text{-Mod})}(\text{Hom}_{A\text{-Mod}}(_AA, -), \mathcal{R})$$

and by the (enriched) Yoneda lemma (cf. 2.5.1) the latter is in bijection with $\mathcal{R}_A A$, i.e. as objects of $\mathbb{K}$-Mod, $E \cong A$. □

Either by the representability claim in the proof or by comparing with the classical statement, we see that $\mathcal{R}$ deserves to be thought of as the regular representation of A-Mod. If A has more structure, such as being a Hopf algebra, Tannaka reconstruction (a 'corollary' of Tannaka duality, discussed above) says that we can recover the algebra from this regular representation of A-Mod, just as $\text{End}_{A\text{-Mod}}(_AA)^{\text{op}} \cong A$ as algebras.

To tie together the two major application areas for representation theory we have examined, namely groups and quivers, we note that there is a construction unifying the group algebra and the path algebra. This is known (perhaps unsurprisingly) as the *category algebra*.

Definition 6.4.8. Let $\mathcal{C}$ be a small category and let R be a ring. The *category algebra* $R[\mathcal{C}]$ is the R-algebra whose underlying R-module is the free module over the set of morphisms of $\mathcal{C}$ and with multiplication defined on basis elements by

$$m(q, p) = \begin{cases} q \circ p & \text{if defined} \\ 0 & \text{otherwise} \end{cases}$$

where p, q are morphisms in $\mathcal{C}$.

Examples 6.4.9.

(a) Let G be a group and $\mathcal{G}$ the 1-object category with object $*$ and morphisms $\mathcal{G}(*, *) = G$. Every morphism in $\mathcal{G}$ is invertible: recall that we call such categories *groupoids*. Then by examining the precursor constructions to Definition 5.2.5, we see that $R[\mathcal{G}] = R[G]$ (where $R[G]$ is the group R-algebra constructed exactly as in the case $R = \mathbb{K}$).

That is, the group algebra is the category algebra of the associated 1-object groupoid.

(b) Let $\mathcal{Q}$ be a quiver and $\mathcal{P}(\mathcal{Q})$ the path category of Definition 2.1.3. Then $R[\mathcal{P}(\mathcal{Q})] = R[\mathcal{Q}]$.

That is, the path algebra is the category algebra of the path category.

Remark 6.4.10. If the category $\mathcal{C}$ actually has finitely many objects and morphisms, the category algebra is canonically a coalgebra (6.3.1) with comultiplication $\Delta \colon R[\mathcal{C}] \to R[\mathcal{C}] \otimes R[\mathcal{C}]$, $\Delta f = f \otimes f$ on morphisms (i.e. every morphism is group-like). In the 1-object groupoid case, this recovers the coalgebra structure on the group algebra of 6.3.5(a). Indeed, this comultiplication is an algebra homomorphism, so we have the beginnings of a Hopf algebra structure; with additional properties (such as hold in the group case) we may be able to make $R[\mathcal{C}]$ a Hopf algebra.

Note that if $\mathcal{C}$ does not have finitely many objects, the category algebra will not be unital: as in the path algebra case, one has a system of orthogonal idempotents, one for each object, coming from the identity morphisms of each object. However, the multiplicative identity would have to be the sum of these and this is not a well-defined element of $R[\mathcal{C}]$ if there are infinitely many objects; we would have to pass to some suitable completion.

So, although we have generally argued in favour of replacing groups and quivers with suitable algebras with the same representation theory, one can argue that we should read the correspondence the other way. Namely, categories allow us to handle non-unital algebras more smoothly.

As we have seen in the two special cases, one would hope for a relationship between linear representations of a category and modules for its category algebra. Indeed, such a theorem is true and the proof is essentially that that we gave for quivers:

Theorem 6.4.11. *Let $\mathcal{C}$ be a small category and $\mathbb{K}$ a field. Then there is an equivalence of categories*

$$\mathrm{Rep}(\mathcal{C}) \cong \mathbb{K}[\mathcal{C}]\text{-Mod}$$

For two algebraic objects (e.g. groups, rings, algebras etc.), the strongest (categorically sensible) form of "sameness" is *isomorphism*. If two objects are isomorphic, in particular their representation theories are the same. Having the same representation theory is known as *Morita equivalence*. A representation theorist would then say that the above theorem states that

> a category and its category algebra are Morita equivalent.

There are even weaker but still useful forms of "sameness", notably *derived equivalence*. However, this seems like a good point to stop.

Bibliography

[AB]   J. L. Alperin and Rowen B. Bell. *Groups and representations*, volume 162 of *Graduate Texts in Mathematics*. Springer-Verlag, New York, 1995.

[AF]   Marlow Anderson and Todd Feil. *A first course in abstract algebra*. CRC Press, Boca Raton, FL, third edition, 2015.

[Alu]   Paolo Aluffi. *Algebra: chapter 0*, volume 104 of *Graduate Studies in Mathematics*. American Mathematical Society, Providence, RI, 2009.

[AM]   Michael F. Atiyah and I. G. Macdonald. *Introduction to commutative algebra*. Westview Press, Boulder, CO, 2016.

[AR]   Maurice Auslander and Idun Reiten. Representation theory of Artin algebras. III. Almost split sequences. *Comm. Algebra*, 3:239–294, 1975.

[ARS]   Maurice Auslander, Idun Reiten, and Sverre O. Smalø. *Representation theory of Artin algebras*, volume 36 of *Camb. Stud. Adv. Math.* Cambridge University Press, Cambridge, 1995.

[ASS]   Ibrahim Assem, Daniel Simson, and Andrzej Skowroński. *Elements of the representation theory of associative algebras. Vol. 1*, volume 65 of *London Mathematical Society Student Texts*. Cambridge University Press, Cambridge, 2006.

[BG]   Ken A. Brown and Ken R. Goodearl. *Lectures on algebraic quantum groups*. Advanced Courses in Mathematics, CRM Barcelona. Birkhäuser Verlag, Basel, 2002.

[Car]   R. W. Carter. *Lie algebras of finite and affine type*, volume 96 of *Cambridge Studies in Advanced Mathematics*. Cambridge University Press, Cambridge, 2005.

[DK]   Yurij A. Drozd and Vladimir V. Kirichenko. *Finite-dimensional algebras*. Springer-Verlag, Berlin, 1994. Translated from the 1980 Russian original and with an appendix by Vlastimil Dlab.

[EGNO] Pavel Etingof, Shlomo Gelaki, Dmitri Nikshych, and Victor Ostrik. *Tensor categories*, volume 205 of *Math. Surv. Monogr.* American Mathematical Society (AMS), Providence, RI, 2015.

[EH] Karin Erdmann and Thorsten Holm. *Algebras and representation theory*. Springer Undergraduate Mathematics Series. Springer, Cham, 2018.

[FH] William Fulton and Joe Harris. *Representation theory*, volume 129 of *Graduate Texts in Mathematics*. Springer-Verlag, New York, 1991.

[GW] Kenneth R. Goodearl and R. B. jun. Warfield. *An introduction to noncommutative Noetherian rings.*, volume 61 of *Lond. Math. Soc. Stud. Texts*. Cambridge University Press, Cambridge, 2nd ed. edition, 2004.

[IM] Kostiantyn Iusenko and John William MacQuarrie. The path algebra as a left adjoint functor. *Algebr. Represent. Theory*, 23(1):33–52, 2020. https://doi.org/10.1007/s10468-018-9836-y.

[Jac] Nathan Jacobson. *Basic algebra. I*. W. H. Freeman and Company, New York, second edition, 1985.

[JL] Gordon James and Martin Liebeck. *Representations and characters of groups*. Cambridge University Press, New York, second edition, 2001.

[Lor] Martin Lorenz. *A tour of representation theory*, volume 193 of *Graduate Studies in Mathematics*. American Mathematical Society, Providence, RI, 2018.

[Maj] Shahn Majid. *A Quantum Groups Primer*, volume 292 of *London Mathematical Society Lecture Note Series*. Cambridge University Press, Cambridge, 2002.

[ML] Saunders Mac Lane. *Categories for the working mathematician*, volume Vol. 5 of *Graduate Texts in Mathematics*. Springer-Verlag, New York-Berlin, 1971.

[Rie] Emily Riehl. *Category theory in context*. Dover Publications, Mineola, NY, 2016.

[Rot] Joseph J. Rotman. *An introduction to the theory of groups*, volume 148 of *Graduate Texts in Mathematics*. Springer-Verlag, New York, fourth edition, 1995.

[Sag] The Sage Developers. *SageMath, the Sage Mathematics Software System (Version 10.6)*, 2025. https://www.sagemath.org.

[Sch] Ralf Schiffler. *Quiver representations*. CMS Books in Mathematics/Ouvrages de Mathématiques de la SMC. Springer, Cham, 2014.

[Sha]   Amit Shah. Krull–Remak–Schmidt decompositions in Hom-finite
          additive categories. *Expo. Math.*, 41(1):220–237, 2023. `https://
          doi.org/10.1016/j.exmath.2022.12.003`.

[ST]   Geoff Smith and Olga Tabachnikova.   *Topics in group the-
          ory*. Springer Undergraduate Mathematics Series. Springer-Verlag
          London Ltd., London, 2000.

[Wei]   Charles A. Weibel. *An introduction to homological algebra*, volume 38
          of *Cambridge Studies in Advanced Mathematics*. Cambridge Univer-
          sity Press, Cambridge, 1994.

Index of SageMath commands

Index

About the Team

Alessandra Tosi was the managing editor for this book.

Annie Hine proof-read this manuscript.

Jeevanjot Kaur Nagpal designed the cover. The cover was produced in In-Design using the Fontin font.

The author typeset the book in LaTeX and compiled the index.

The conversion to the HTML edition was performed with latexml, an open-access tool freely available at https://math.nist.gov/~BMiller/LaTeXML/.

Hannah Shakespeare was in charge of marketing.

This book was peer-reviewed by Billy Woods (Pathways Department, University of Essex), Matthew Pressland (Laboratoire de Mathématiques Nicolas Oresme, Université de Caen-Normandie) and an anonymous referee. Experts in their field, these readers give their time freely to help ensure the academic rigour of our books. We are grateful for their generous and invaluable contributions.